Generis
PUBLISHING

Nature, Humanity, and the Fragile Balance of Existence

Mustafa H. Aydoğdu

Title: **Nature, Humanity, and the Fragile Balance of Existence**

ISBN: 979-8-89966-348-2

Author: Mustafa H. Aydoğdu

Cover image: www.pixabay.com

Publisher: Generis Publishing
Online orders: www.generis-publishing.com
Contact email: info@generis-publishing.com

Nature, Humanity, and the Fragile Balance of Existence

(A Philosophical Inquiry into the Ethical, Ecological, and Existential Dimensions of Nature)

Table of Contents

About the Author

Prof. Dr. Mustafa Hakkı Aydoğdu

Prof. Dr. Mustafa Hakkı Aydoğdu is an academic who approaches the relationship between nature and humanity not only from an academic standpoint but also through a field-based and conscientious perspective.

Born in Türkiye, after graduating from the Middle East Technical University, Faculty of Engineering, Department of Civil Engineering, he actively participated in the GAP–Harran Plain irrigation projects. The practical experience he gained in this field later formed the foundation of his academic pursuits.

He subsequently joined the Southeastern Anatolia Project (GAP) Regional Development Administration, where he assumed various technical and administrative roles, including responsibilities at the level of regional director. During this period, he also completed a second bachelor's degree in Public Administration, thereby building a bridge between engineering and public policy.

Continuing his academic journey at Harran University, Aydoğdu served in multiple capacities including faculty member, administrator, and Secretary General. He earned his master's degree in Civil Engineering with a focus on water resources, and later completed his Ph.D. in Agricultural Economics, specializing in natural resources and environmental economics.

He currently serves as a Professor in the Department of Agricultural Economics, Division of Agricultural Policy and Extension at the Faculty of Agriculture, Harran University, and also works as a consultant at the General Directorate of Combating Desertification and Erosion under the Ministry of Environment, Urbanization and Climate Change.

His areas of expertise include natural resources and environmental economics, land use planning, agricultural irrigation, water policies, and water pricing. He has contributed to numerous national and international projects as a principal investigator, researcher, and consultant.

This book is not only the voice of an academic, but also the expression of an individual who strives to defend the rights of nature and natural resources, uphold humanity's responsibility, and preserve hope for future generations.

About the Book

The purpose of this book is to reveal the roots and current state of ecological crises by examining the relationship between humankind and nature—bestowed upon humanity as a vital resource—from theological, historical, philosophical, and ethical perspectives. It also aims to shed light on the fragility of nature and the degradation processes shaped by human interventions.

While discussing how humanity's perception of nature has evolved within a broad spectrum—from systems of belief to ethical theories—the book analyzes the dimensions of issues such as climate change, global warming, the water crisis, erosion, desertification, and land degradation that directly affect human life. Each chapter questions the human–nature relationship through philosophical thought, ethical responsibility, and theological and cultural approaches, providing a snapshot of the current global and national situation.

Target Audience of the Book:

- Policymakers, decision-makers, academics, and researchers working on ecology, environmental philosophy, and sustainability,
- Educators, students, and civil society representatives who generate ideas on environmental awareness and the nature–ethics relationship,
- And all curious minds seeking to understand the holistic dimensions of ecological crises.

This book demonstrates that the relationship between humanity and nature is not merely a technical matter but also an ethical, philosophical, and cultural one. It emphasizes that to comprehend these crises, one must look not only at scientific data but also at systems of value. Thus, it seeks to make the "silent cry" of nature heard and confront humanity with its own responsibility.

Although each chapter is designed to stand independently, they are all structured to serve a common purpose. Short repetitions appearing throughout the book are consciously preserved to ensure thematic integrity and deepen the reader's understanding. During the preparation process, an extensive literature review was conducted; even studies that did not directly appear in the text were cited in the chapter references out of respect for their intellectual contribution. Due to the diversity of sources, some numerical discrepancies may occur—these arise from methodological differences among datasets and research works.

Artificial intelligence technologies were utilized in the book's cover design and content layout. This book does not claim to be flawless. On the contrary, it is not an invitation to perfection, but to questioning and intellectual exploration. Its primary purpose is to encourage readers to see their relationship with nature from a new perspective—to perceive it as an integral framework shaping the meaning and boundaries of human life. The main expectation is that, while enjoying the reading process, the reader will also reflect on the value of the resources they possess and deepen their sense of responsibility. If this book succeeds in awakening both intellectual curiosity and an opportunity to reevaluate these resources with care and humility, then it will have achieved its purpose.

Prof. Dr. Mustafa H. AYDOĞDU

Acknowledgments

This book is not merely a written text; it is a reflection of my profound sense of responsibility toward the Creator, all created beings, nature, soil, water, and life itself. Every line has been inspired by my experiences—from the shadow of a tree, from the silence of a handful of soil, from the journey of a single drop of water.

First and foremost,

To those who live by and take to heart Yunus Emre's words: "Cherish the created for the sake of the Creator" and "Love all beings for the sake of the One who created them."

To those who see a tree, a river, a bird, or a mountain as sacred because they are creations of the Divine;

To those who protect nature, defend the environment, and struggle for natural resources;

To everyone whose words, actions, hearts, and even names carry the essence of "Nature";

To those who not only speak but take action and open new paths—

Your effort, your voice, and your hope echo in every word of this book.

I extend my deepest gratitude to all volunteers, scientists, educators, writers, civil society organizations, workers, farmers, citizens, and countless unseen heroes who listen to nature's silent cries and defend its rights. Thanks to you, life on this Earth remains possible.

To my beloved wife, children, and family—your unwavering love, patience, and understanding have always been my greatest strength. With heartfelt gratitude to my roots, with whom I have learned the true meaning of life and existence.

And to my colleagues and companions in thought, labor, and hope—your ideas, support, critiques, and solidarity form the foundation stones of this work. I am deeply grateful to each of you.

Finally, this book is dedicated to all those whose breath fills the spaces between its lines—to every heart and soul who continues, with the faith of the rising sun, to defend the environment, nature, and natural resources.

Let us not forget:

One day, nature will prevail...

And on that day, a new era for humanity will begin.

With endless gratitude and love,

Prof. Dr. Mustafa H. AYDOĞDU

Foreword

This book was born out of an urgent need — not merely to understand the crises of nature, but to *listen* to them. The human journey has long been celebrated as a triumph of intellect and will. Yet in our pursuit of mastery, we have drifted away from the rhythm of existence that once held us in harmony with the Earth.

Nature, Humanity, and the Fragile Balance of Existence (A Philosophical Inquiry into the Ethical, Ecological, and Existential Dimensions of Nature) invites the reader to revisit that forgotten dialogue — between the sacred and the scientific, between the ethical and the ecological. It is not merely a book about nature's decline, but a reflection on the moral landscape of human thought, responsibility, and belonging.

Each chapter traces the fragile thread that connects the human spirit to the natural world — from the silence of the soil to the roar of storms, from the waters that sustain life to the deserts that warn us of its limits. This is not a call to despair but to remembrance — to rediscover what it means to be human *within* nature, not *above* it.

This book tells the story of beings that once remained silent but now cry out — the environment, nature, and natural resources. The soil, water, air, and all living creatures that have long offered unconditional hospitality to communities, civilizations, faiths, and cultures have been neglected for too long. They have been dominated in the name of progress and modernity. Yet nature and its resources are no longer mere objects — they have become *subjects* that speak. They warn us, remind us, and call us to account.

The world was not created at random; it was brought into existence with balance, order, and purpose — endowed with the means necessary for the survival of humans and all living beings. This harmony reflects the power and grace of God, the true owner and sustainer of all existence.

Today, the collapse of ecosystems is not merely an environmental crisis; it mirrors a profound fracture across belief, morality, politics, and philosophy. The scientific community has revealed the causes of ongoing issues such as climate change, desertification, water scarcity, and biodiversity loss. Yet this book seeks to go beyond: it questions the very foundation of the human–nature relationship. It proposes a new path — one not built upon economic interest but upon faith and moral responsibility, equity, respect, and a consciousness of coexistence.

Each chapter offers not only technical analysis but also a moral awakening — a call of conscience. This call does not merely seek solutions; it invites the reader to re-establish humanity's relationship with nature. For in striving to save nature, we are, in truth, striving to rebuild our own humanity.

"Nature is not merely a resource — it is being itself. Within being lies humanity; within humanity lies nature." In most studies concerning the environment, nature, and natural resources, a few questions dominate the discourse: *"How much is left?", "Who uses how much?", "For what purpose?", "What is its economic value?", "How should it be priced?"*

This book, however, takes a step back and asks a more fundamental question: **"What is value — and for whom?"**

Once sacred, soil and water were commodified in the "modern" era and later redefined as heritage to be preserved. Today, we stand at a new threshold: we can neither return to old impulses of domination nor remain content with nostalgic conservation. We must decide now — will we become a civilization that future generations will question for consuming nature, or one remembered for learning to live in harmony with it?

The aim of this book is not merely to inform, but to awaken — a question in the reader's mind and a sense of responsibility in their heart. The essential matter is this: **to succeed in living together.**

What is written here is not merely a warning — it is an invitation. An invitation to morality, to faith, to science, to justice, and to the culture of coexistence.

> "In every grain of soil, in every drop of water, and in every breath of wind, nature whispers the story of balance — a story we once knew by heart."

With greetings, respect, and affection,

Prof. Dr. Mustafa Hakkı AYDOĞDU

CONTENTS

3rd CHAPTER

NATURE'S REBELLION: THE MULTIPLICITY OF CRISES

(The Anatomy of Ecological Crises: What Has Happened and What Is Happening?; Causes: Humanity's Anomalous Interaction with the Atmosphere; The Balance of the Planet Is Being Shaken; Paris Agreement and Country-Based Commitment Comparisons; Global Warming and Glacial Mass Loss; Extreme Weather Events: Frequency, Intensity, and Climate Linkages; Agriculture and Food Crisis: A Global Threat Deepened by Climate Change; Migration and Security: In the Context of Climate Change)

4th CHAPTER

WATER RESOURCES, NATURAL SYSTEMS, AND ETHICAL RESPONSIBILITY: A HOLISTIC ASSESSMENT FROM GLOBAL AND TURKISH PERSPECTIVES

(Water: The Heart of Natural Existence and Moral Responsibility; Global Analysis of Precipitation, Drought, and Climate Change; Drought and Its Classifications; the Global Status of Water Resources; Türkiye's Water Assets and Utilization; Sectoral Evaluation of the Water Problem; General Recommendations for Sectoral Water Use; An Overview of Agricultural Irrigation Systems; Global Sectoral Water Efficiency and the Economic Value of Water; Sectoral Water Efficiency and Economic Value of Water in Türkiye; Natural Resources, Water, and the Ethics of Scarcity: Between Moderation, Responsibility, and Justice; The Ethics of Scarcity in Beliefs and Philosophical Systems; A Ten-Step Roadmap for Sustainable Water Management)

5th CHAPTER

EROSION AND NATURE: THE ANATOMY OF A SILENT EXHAUSTION

(The Silent Cry of the Soil; Islamic, Philosophical, and Cultural Perspectives: The Memory of the Soil; Definition and Types of Erosion; Causes of Erosion; Ecological Impacts of Erosion; Socioeconomic Impacts of Erosion; Erosion in Türkiye: Situation Analysis and Statistical Overview; Global Perspective: Erosion, Climate Change, and Losses; Ten-Step Proposal for Sustainable Soil Management)

6th CHAPTER

DESERTIFICATION AS A GLOBAL THREAT

(The Silent Exhaustion of the Soil: A Multidimensional Analysis of Desertification; The Social Justice Dimension of Desertification; The Reality of Desertification in Türkiye: Trends and Statistics; International Approaches to Combating Desertification; The Impacts of Desertification on the Global Economy; Methods for Combating Desertification – Nature-Based Solutions and Sustainable Land Management; Policy and Strategic Roadmap Recommendations)

7th CHAPTER

LAND DEGRADATION: A GLOBAL THREAT TO THE FRAGILITY OF NATURE AND THE "BALANCING" APPROACH

(The Necessity of Harmony with Nature; Global and Local Perspectives: The Anatomy of Degradation in Numbers; Land Degradation Neutrality Approach: Philosophy, Principles, and Implementation Hierarchy; Türkiye and Land Degradation Neutrality: Target Process and Challenges; Evolution of the LDN Concept in Light of United Nations Models: 1994 – 2007 – 2015; Land Rights, Justice, and Ecological Ethics: A Philosophical Perspective; Sustainable Solutions and New Policy Directions: Adaptation, Restoration, and Participation)

8th CHAPTER

SAND AND DUST STORMS: THE SILENT REBELLION OF NATURE

(A Letter from the Dust; Nature's Reaction: Crossing Ecological Thresholds; The Memory of Dust: Forgotten Balances in Ecosystems; Sand and Dust Storms: Causes, Significance, Impacts, Measures, and Future Scenarios; Reestablishing Dialogue with Nature: Harmonious Lifestyles and Restorative Ecology)

INTRODUCTION

CONCEPTUAL FRAMEWORKS

On the Inclusiveness of the Concept of Nature

The concept of nature represents a sense of unity at philosophical, ecological, and political levels. Nature encompasses all living and non-living entities, the reciprocal relationships among them, and the ecological processes that sustain these interactions. In this respect, nature can be regarded as a higher-order system that includes both the environment and natural resources. The environment expresses the sphere of interaction between humans and nature, while natural resources represent the tangible components of nature that can be utilized by humans.

Thus, nature is not merely a biophysical reality; it is also a multilayered concept situated at the heart of human values, ethical inquiry, and the understanding of ecological justice. This holistic perspective forms the foundation of contemporary sustainability debates (Naess, 1973; Leopold, 1949; Daily, 1997). In the modern era—shaped by human activities—nature is perceived not only as a physical realm but also as a social, cultural, and ethical frame of reference.

The Conceptual Relationship between "Nature," "Natural Resources," and "Environment": Different Faces of the Same Reality

Throughout this book, all three concepts are used, yet in the academic sense, they are closely related but not identical.

Nature: The broadest concept, encompassing all living and non-living beings, processes, and systems that exist independently of or prior to human intervention. Nature includes ecosystems, climate, geological formations, living organisms, the atmosphere, water, and soil as an integrated whole. It does not exclude humanity but rather embraces it as part of its cycle. However, when humans forget their place within this whole, they begin to see nature as "the other."

Natural Resources: The elements of nature that can be utilized for human benefit, forming a subset of nature. In this sense, they represent an instrumental reflection of it, including water, soil, forests, minerals, and energy sources. However, placing human benefit at the center of this concept risks commodifying and depleting nature (Norton, 2005).

Environment: The sphere of interaction between humans and nature, gaining meaning through human presence. Hence, the environment is not merely a physical entity but also an ethical space of relationship.

Nature is the essence of existence; the environment is our relationship with it; and natural resources are the instrumentalized form of this relationship. Therefore, saying "nature" often includes both natural resources and the environment, though these concepts may need to be highlighted separately depending on context.

For example, the expression "protection of nature" implicitly refers to the preservation of natural resources, ecosystems, and the environment. Yet, in policy contexts, terms such as "natural resource management" or "environmental policies" are more technical and field-specific.

Below is a table illustrating the philosophical, ecological, and political distinctions among the concepts of nature, natural resources, and environment:

Dimension	Nature	Natural Resources	Environment
Definition	The entirety of all living and non-living beings, processes, and relationships that exist independently of human intervention.	Components of nature that carry economic, social, or cultural value and can be used for human benefit.	The sphere of interaction between humans and nature—comprising physical, biological, social, and cultural elements.
Philosophical Meaning	Regarded as the essence of existence, the source of life, and the "whole" beyond humanity; it has an ontological foundation.	Holds an "instrumental value," selected and defined by humans according to their needs.	Represents the form of relationship humans establish with nature and the ethical limits of that relationship.
Ecological Meaning	Relates to the dynamic balance of ecosystems, energy cycles, biodiversity, and ecological integrity.	Evaluated as limited and renewable/non-renewable elements within the sustainability of ecosystems.	A systemic concept examining the interaction between natural and artificial elements.

Political Meaning	Possesses a normative value in the context of nature conservation, ecological justice, and climate ethics.	Central to discussions of economic development, resource management, energy policy, and sustainable use.	Forms the foundation of environmental policies, pollution control, urbanization, and sustainable living strategies.
View of Humanity	Sees humans as part of nature.	Positions humans as "users" of nature.	Considers humans as "actors" in continuous interaction with nature.
Ethical Dimension	Serves as the primary reference for deep ecology and ecocentric ethics.	Emphasizes utilitarian and management-oriented approaches.	Associated with principles of sustainability, environmental justice, and social responsibility.

Ethics and Morality: The Mirror of Thought and Behavior

The nature of the relationship that humanity establishes with the natural world determines its ethical and moral stance. Although ethics and morality are often used interchangeably, they carry distinct layers of meaning in philosophical discourse.

Morality: Morality refers to the normative set of values that guide the behavior of individuals and societies, defining distinctions between "good and evil" or "right and wrong" (Aristotle, 4th century BCE/2009). In this sense, morality is shaped by culture, molded through social experience, and embodied in individual actions. It is formed by the collective norms of society and manifested through conduct. Morality regulates how humans should behave in daily life—it is society's collective answer to the question, "How should we act?"

Ethics: Ethics, on the other hand, is the philosophical discipline that questions the origins, validity, and universality of moral norms (Kant, 1785/2011). Ethics is less concerned with individual actions than with the theoretical analysis of moral systems. Whereas morality asks, "How should we act?", ethics asks, "Why should we act that way?" Ethics is the intellectual depth of morality.

Ethics seeks answers to the why and how questions underlying moral behavior, aiming to ground moral conduct not only in social acceptance but also in rational justification. In this sense, ethics constructs the theoretical framework of morality and evaluates it through a critical lens.

While ethics analyzes the intentions, values, and consequences behind individual actions, morality defines the socially accepted forms of those actions. For instance, the statement “Lying is bad” is a moral judgment; whereas questions such as “Why is lying wrong?” or “Under what circumstances might lying be acceptable?” belong to the domain of ethics.

Therefore, morality pertains to behavioral reality, while ethics is the field of intellectual reflection. This relationship can be summarized as follows:

“Ethics is the philosophy of morality; morality is ethics embodied in life. Morality concerns action, ethics concerns thought.”

Habermas (1991) treats ethics and morality as two levels of communicative reason: ethics as the sphere of individual self-reflection, and morality as the product of social consensus. Thus, every ethical stance is simultaneously a call to social responsibility.

Differences between Ethics and Morality

Criterion	Ethics	Morality
Definition	The philosophical discipline that studies moral principles, values, and behavior.	The set of social or individual beliefs about what is “right–wrong” or “good–bad.”
Level	Theoretical (conceptual) level.	Practical (applied) level.
Purpose	To justify and question why moral actions are right or wrong.	To determine and guide how people should behave.
Type of Question	“Why is it right?”, “According to what is it good?”, “Are there universal principles?”	“What is right?”, “Is this behavior acceptable?”
Source	Reason, philosophical thought, and critical analysis.	Society, tradition, religion, culture, and family values.
Example	“How do we explain why lying is wrong?”	“Lying is bad.”
Field	Philosophy (particularly moral philosophy).	Social life, individual behavior, cultural norms.
Relationship	Reflects on, analyzes, and questions morality.	The phenomenon studied and interpreted by ethics.

The Ethics and Morality of Nature

Every human act toward nature—whether the destruction of a forest or the protection of a river—embodies an ethical judgment. Cutting a tree, polluting water, or exhausting the soil is not merely a technical act; it is an existential attitude. Every action directed at nature is, in essence, an action directed toward existence itself. This understanding transforms the protection of nature from a mere "duty" into a way of being.

To harm nature is not only to harm ecosystems but also to damage the inner balance of the human self—disrupting the harmony within one's own being. Ethics, in this context, emerges as a consciousness of responsibility toward nature, whereas morality represents the embodied form of that consciousness in action.

In this sense, ethics is the intellectual awareness of responsibility toward nature, and morality is that awareness translated into behavior. The modern pursuit of sustainability, in fact, rests upon a profound ethical question:

"Where does the human right over nature end, and where does responsibility begin?"

To understand the distinction among nature, environment, and natural resources correctly is to determine the direction of the answer to this question. Nature represents wholeness, the environment represents interaction, and natural resources represent use. Ethics serves as the intellectual framework that balances these three dimensions; morality is the embodiment of that balance in practice.

"Without ethical thought, nature cannot be protected; without morality, ethics cannot be lived."

Protecting nature is not achieved by merely loving it, but by remembering that we belong to it. Nature is not an object external to us—it is the very essence of our existence.

References

Aristoteles. (2009). Nicomachean ethics (W. D. Ross, Trans.). Oxford University Press. (Original work published ca. 4th century BCE)

Carson, R. (1962). Silent Spring. Houghton Mifflin.

Daily, G. C. (Ed.). (1997). Nature's Services: Societal Dependence on Natural Ecosystems. Island Press.

Habermas, J. (1991). Moral consciousness and communicative action (C. Lenhardt & S. W. Nicholsen, Trans.). MIT Press.

Kant, I. (2011). Groundwork of the metaphysics of morals (M. Gregor, Trans.). Cambridge University Press. (Original work published 1785)

Leopold, A. (1949). A Sand County Almanac and Sketches Here and There. Oxford University Press.

Mill, J. S. (1998). Utilitarianism. Oxford University Press. (Original work published 1863)

Naess, A. (1973). The shallow and the deep, long-range ecology movement: A summary. Inquiry, 16(1–4), 95–100. DOI: 10.1080/00201747308601682

Norton, B. G. (2005). Sustainability: A Philosophy of Adaptive Ecosystem Management. University of Chicago Press.

Pepper, D. (1984). The Roots of Modern Environmentalism. Routledge.

1st CHAPTER

HUMAN AND NATURE: THE PHILOSOPHY OF HARMONY AND THE PRICE OF DISHARMONY

Chapter Summary

This chapter emphasizes that nature should not be perceived merely as a physical entity but as a reflection of the Creator, the sacred, the divine order, existence itself, and moral responsibility. It explores the approaches of various belief systems—such as Islam, Christianity, Judaism, Hinduism, Buddhism, Taoism, Shintoism, and Indigenous and Shamanic traditions—toward nature and natural resources. Each of these traditions views nature and its resources as a sacred trust, a spiritual being, or a realm of moral responsibility.

The intersections between modern philosophical and ethical perspectives (such as anthropocentrism, ecocentrism, deep ecology, and virtue ethics) and these belief systems are analyzed. This intersection elevates the human–nature relationship from a merely instrumental to a moral and existential dimension. Nature is defined as "the silent wholeness of being," offering a multilayered interpretation of human–nature relations through both Western and Eastern philosophical lenses.

The chapter also addresses the internal functioning of ecosystems and the disruption of natural balance due to human intervention, arguing that nature's silence is, in fact, a form of resistance, equilibrium, and wisdom.

Keywords: Nature–Sacred Relationship, Theological and Ethical Approaches, Faith Systems and the Environment, Ecosystem Balance and Silence, Philosophical Ecology

Nature as a Reflection of the Sacred in the Perspective of Beliefs and Religions

For humankind, nature has never been merely a habitat; it has been the reflection of the sacred, the manifestation of divine order, a spiritual guide, and at times, the very object of worship. Across different religious traditions and faiths, perspectives on nature and natural resources are shaped by how each belief system constructs its spiritual relationship with the Creator or the divine. Although approaches to nature differ between those who believe in God and those who do not, all share common teachings of reverence and protection.

In Islamic thought, nature is viewed as a mirror reflecting the existence, power, and wisdom of God. The natural world — the heavens, the earth, the mountains, the winds, the waters, and all living beings — are not merely elements created for human use, but signs that invite reflection on the meaning of existence itself. According to Islamic teachings, human beings are not absolute rulers over nature; rather, they are trustees (stewards) who are obliged to live in harmony with it and protect it.

Within this understanding, the protection of the environment and natural resources is not only an ecological duty but also a moral and spiritual obligation. Wastefulness — understood as the excessive or unnecessary use of resources — is prohibited both ethically and religiously. The fair, balanced, and sustainable use of natural resources is considered a fundamental aspect of humanity's divine purpose on Earth — the fulfillment of its stewardship (khilafah) role.

Islam's environmental philosophy is grounded in the principles of trust (amanah), balance (mizan), and compassion (rahmah). In this context, harming nature is seen not only as an ecological degradation but also as a violation of a sacred trust. Thus, Islam promotes a holistic environmental consciousness that defines humanity's relationship with nature as both a spiritual responsibility and an ethical commitment (Ardoğan, 2012).

In the Bible, nature is described as a system that God created "good" and entrusted to humanity (Genesis 1:31). "The heavens declare the glory of God" (Psalms 19:1). Catholic social teaching emphasizes that humans must build a compassionate relationship with nature (Laudato Si', 2015). Pope Francis's ecological call reinvigorated Christian environmental ethics, framing nature as both God's creation and the embodiment of divine glory.

In the Jewish scriptures (Tanakh), nature belongs to God, and humankind must use it responsibly (Leviticus 25:23). The concept of the Shabbat Year (Shmita), during which land is left to rest every seventh year, embodies an ecological ethic of renewal and respect (Leviticus 25:1–7).

In Hinduism, elements of nature are manifestations of divine beings. The Ganges River is revered as a sacred mother, cows are holy beings, and trees are the dwelling places of the god Vishnu. For Hindus, the gods dwell within nature—they are the divine face of nature. According to the principle of Ahimsa (non-violence), harming nature generates negative karma; thus, nature is not only protected but also worshipped.

In Buddhism, every sentient being experiences dukkha (suffering), and harming nature contradicts karuṇā (compassion). Buddhism is built upon harmony with nature and shared suffering. Taoism and Shintoism, on the other hand, emphasize unity with the spirit of nature. Tao represents the way of nature—its flow and essence. To resist nature is to resist the Tao. "Man follows the Earth, Earth follows Heaven, Heaven follows the Tao, and the Tao follows what is natural" (Tao Te Ching). In Shintoism, the Kami (spirits) inhabit mountains, forests, and rivers; every natural formation is sacred and ensouled.

In Indigenous and Shamanic beliefs, nature represents a sacred cycle of life. According to the Mother Earth principle—such as among Native American traditions—nature is a living mother. In Shamanism, every entity in nature possesses a spirit (animism). Communication with nature occurs through ritual and inner harmony.

The spiritual bond between humanity and nature in many belief systems is not merely an abstract teaching but a lived practice. For example, Buddhist monks in the Chiang Mai region of Thailand conduct silent meditations within forest monasteries, believing that nature itself is the teacher of silence (Sponsel, 2012). In Shamanic rituals, thanking the earth after a hunt or acknowledging the spirit of each plant symbolizes a dialogue with nature. Similarly, Shinto shrines in Japan are sacred spaces inhabited by kami—spirits of mountains and rivers. These examples illustrate that nature is perceived as a temple, a teacher, and a companion. Nature, therefore, is not simply an entity; it is a living, teaching, and feeling force.

In all belief systems, nature is the silent voice of the sacred. Across religions, nature and natural resources are viewed as manifestations of creative power, as sacred spaces, moral beings, or spiritual wholes. To harm nature is not merely an environmental violation but a theological, moral, and ethical deviation. Humanity is both a part of and a guardian of this sacred unity. All faiths root this guardianship in love, respect, responsibility, gratitude, and humility.

Shared Ethical Principles toward Nature in Different Faith Traditions

What do various faiths share in common? What are the core ethical values that different religions uphold regarding nature and natural resources? Across diverse spiritual traditions, there exists a shared ethical foundation rooted in reverence, gratitude, and responsibility toward nature. The following table summarizes the main principles and their reflections within different faiths:

Core Principle	Explanation	Religious Correspondence
Trust and Stewardship	Nature and natural resources must be protected as divine trusts entrusted by the Creator.	*Islam:* Humanity is steward and resources are divine trusts (Qur'an, Al-An'ām 6:165). *Judaism:* The land belongs to God (Leviticus 25:23).
Humility	Humanity must approach nature with humility rather than arrogance.	*Buddhism: Karuṇā* (compassion) and mindfulness; *Taoism:* Submission to the Tao; *Christianity:* The spirit of servanthood.
Compassion	All beings deserve care and should not be harmed.	*Hinduism: Ahimsa* (non-violence); *Islam:* Mercy toward all creatures; *Christianity:* "Love your neighbor" (Matthew 22:39).
Cyclicity	Respecting the rhythm of nature—allowing time for renewal and rest.	*Judaism:* The *Shabbat Year* (Leviticus 25); *Indigenous Traditions:* Mother Earth principle and the sacred cycle of life.
Gratitude and Prayer	Expressing inner connection and thankfulness for nature's blessings.	*Shamanism:* Post-hunt prayers of gratitude; *Islam:* Thankfulness for divine bounty; *Christianity:* Prayer and thanksgiving.

The principles above—trust, humility, compassion, cyclicity, gratitude, and prayer—represent a universal ethical and spiritual framework that humanity has sought throughout history. These principles have been incorporated into various forms in different religions, cultures, and teachings. A sense of trust is central to humankind's relationship with nature and natural resources.

In Islamic thought, the human being exists on Earth not as a possessor but as a trustee—one who bears responsibility rather than ownership. This worldview regards nature not as a domain open to unlimited human exploitation, but as a divine trust that must be protected, respected, and used with justice and moderation. Humanity's relationship with nature is therefore defined not by dominance, but by responsibility and moral balance.

Within this framework, humility signifies the awareness of one's place within the cosmic order and the acknowledgment of human limitation before the Creator. Arrogance, by contrast, is viewed not only as a moral flaw but as a disruption of both the natural order and social harmony. At the heart of Islamic ethics lies compassion (rahmah) — a virtue that extends beyond human beings to encompass all living creatures, the soil, the waters, and even the trees. Compassion represents the recognition of life's sanctity and the commitment to preserving the integrity of creation.

The principle of cyclicality in Islam carries both ecological and metaphysical meaning. The cycle of death and rebirth reflects not only the continuity of nature but also the eternal journey of the human soul. The seed that is buried in the earth and rises again symbolizes both the renewal of life and the resurrection that transcends the material world.

Gratitude (shukr) and prayer (du'a) form the spiritual language of humanity's relationship with nature. Gratitude, in this sense, is more than mere thankfulness; it is a form of awareness—a conscious alignment with the order of creation. Such consciousness opens the way to both spiritual harmony and ecological balance.

Despite the deeply sacred connections that faith systems have historically upheld toward nature, the "modern" human has grown increasingly alienated from it—forgetting its sanctity, cyclicity, and wisdom. Today's environmental crises are not only technological or political but also spiritual and moral in essence.

"The human being who has distanced himself from nature has not only lost the natural world but also the depth of his own faith."

This loss has deepened with the erosion of virtues such as reverence, gratitude, and humility toward the natural world. Contemporary environmental crises—climate change, species extinction, soil degradation—should thus be read not only as scientific phenomena but also as spiritual warnings. Humanity must once again approach nature not merely with knowledge but with faith, ethics, and sensitivity.

In all belief systems, nature and natural resources are given not simply to be seen or consumed, but to be felt, lived, and sustained. They are divine trusts. As Yunus Emre beautifully expressed:

"Love all creation for the sake of the Creator."

All beings in the universe share the same Creator, who is the essence and the greatest of all. Therefore, we must not only see with our physical eyes but also perceive with the eye of the heart—learning to love and protect.

Nature: The Silent Wholeness of Being

The expression "the silent wholeness of being" aims to help us perceive nature not merely as a collection of physical entities but as a plane of life — a holistic system of balance intricately intertwined with humankind and all living beings. This metaphor invites us to see nature and natural resources not as consumable or observable systems but as a living sphere that encompasses humanity, embraces other beings, and sustains its own rhythms through divine design.

This approach compels us to reposition nature and natural resources within philosophical, scientific, and socio-political contexts. Nature is not merely a resource but the essence of life itself — a silent yet ever-teaching mentor, a master of balance and generosity. Nature and natural resources predate humankind; they are systems sustained by the Creator's wisdom, functioning through intrinsic laws and cycles. Ecosystems, species, climate, soil, water, air, and vitality — these are not merely components of nature; they are nature itself. Nature is neither a resource nor a backdrop; it is a living totality that nourishes, transforms, and sustains all existence.

By divine decree, everything in nature exists in measure. Nothing is created in excess; everything is fashioned with purpose. In this sense, nature possesses — through its Creator — what may be called a "measured intelligence", a harmony of proportion and purpose (Leopold, 1949). Its sustainability lies in reciprocity and equilibrium.

In Western philosophy, nature has often been positioned as the opposite of human reason (Descartes, 1641), whereas in Eastern philosophy, the focus is on living in harmony with it. Spinoza's Deus sive Natura ("God or Nature") conceives of the divine not as transcendent but as immanent within nature (Spinoza, 1677). To Spinoza, nature is self-generating, self-sustaining, and governed by its own laws — requiring no external intervention.

Heidegger (1954) rejects the notion of nature as a mere "tool" or "object" to be mastered; instead, he defines it as "the openness of being itself." Nature, in this sense, is not a measurable instrument but a realm of revelation — an existence that unfolds rather than serves. His metaphor of the "totality of tools" reminds us that nature should not be understood through materialist or utilitarian logic, but through reverence, justice, and silence.

In Eastern philosophy — particularly Taoism — nature (Tao) represents the very flow of life. Lao Tzu (6th century BCE) teaches that "nature does not speak, yet accomplishes all things." This encapsulates the essence of silent wholeness: the power of stillness, harmony, and non-coercion.

The Abrahamic faiths, regard nature not merely as an observable phenomenon but as the manifestation of divine wisdom. Perfection in this context lies not in nature's supposed flawlessness but in humanity's fidelity to the divine balance established by the Creator. Ultimately, the perception of perfection in nature across the revealed religions is shaped by this reciprocal bond: nature is not humanity's dominion but a sacred companion in existence.

This understanding — both ethical and ecological — offers a means of restoring the "sense of measure" that modern civilization has lost. To live in harmony with nature, therefore, is not merely an ecological necessity but a spiritual act of justice, humility, and remembrance.

The Intersection of Theological Approaches and Ethical Theories on Nature and Natural Resources: The Sacred, Responsibility, and the Morality of Life

Across many faith traditions, nature and natural resources are perceived as manifestations of the Creator, divine order, wisdom, and transcendence. Modern ethical theories, meanwhile, interpret nature either as an instrument serving human purposes or as an entity possessing intrinsic value. The convergence of these two perspectives deepens our understanding of the philosophical and theological foundations of sustainability, environmental justice, and the human–nature relationship — revealing the possibility of moral harmony between faith and reason.

In Islamic thought, nature is perceived as a reflection of divine order and wisdom. In the sacred revelation, nature is not an object to be dominated, but a sign (ayah) to be contemplated and understood. From this perspective, the human being holds no right to harm nature; rather, humanity is regarded as a trustee and guardian, entrusted with the responsibility to preserve and protect the natural world and its resources. This Islamic perspective aligns not with neutral utilitarianism or deontological ethics but with an ethic of trust and responsibility (Denny, 1998).

In Catholic environmental theology (for example, Pope Francis's Laudato Si'), nature is described as a "common home." Human authority over creation is not absolute; it is bounded by the moral duty of care. This creates a bridge between the theological model of "sacred stewardship" and the ethical framework of ecological deontology (White, 1967).

Eastern traditions such as Taoism, Hinduism, and Buddhism view nature as a totality that includes humanity. Humanity does not dominate nature but lives in harmony with it — an approach that underpins ecocentric ethics and deep ecology (Naess, 1989).

Ethical Approaches to Nature

1. Anthropocentrism (Human-Centered Ethics)

Nature and natural resources exist solely to meet human needs. In theological contexts, traditional Western interpretations have often reinforced this view (Genesis 1:28). This anthropocentric stance has been widely criticized as a root cause of modern environmental crises (White, 1967).

2. Theocentric Environmental Ethics

A God-centered environmental ethic in which nature belongs to God and must be preserved out of reverence for the Creator (Attfield, 1991). "The land must not be sold permanently, because the land is mine" (Leviticus 25:23). Here, divine responsibility transforms into ecological responsibility.

3. Ecocentrism and Deep Ecology

All components of nature possess intrinsic value. Arne Naess (1989) emphasized identification with nature and recognition of its inner worth. This view aligns closely with Eastern philosophies such as Tao, Dharma, and Ahimsa, which teach unity, balance, and non-harm toward all life forms.

4. Virtue Ethics and Ecological Virtues

Originating from Aristotle, this approach emphasizes moral excellence in human behavior toward nature and resources — virtues such as moderation, humility, respect, and gratitude. Religions provide theological grounding for these virtues, encouraging attitudes of reverence and balance toward creation.

Philosophical Ecology and the Deep Harmony of Faith

Arne Naess's theory of deep ecology advocates identification with nature and equality among species. This resonates with Taoism's teaching of living in rhythm with the Tao. Alfred North Whitehead's process philosophy views nature not as static but as a continuous process of becoming, aligning with evolutionary theology (Cobb & Griffin, 1976).

The ecological crisis of our age compels the reexamination of traditional human-centered frameworks. Yet, all religions already offer a rich normative language capable of redefining our relationship with nature:

Islam: Trust (amānah) and humility

Christianity: Compassion and care

Buddhism: Non-harm and empathy

Hinduism: Harmony and karma

Indigenous Traditions: Gratitude and cyclicity

Each of these values can translate into the ethical principles of sustainability, responsibility, morality, and justice. Nature, therefore, is not merely an environment we inhabit but a moral being — a partner in the sacred covenant of existence.

Religious traditions, by sanctifying nature, provide the moral foundation upon which ethical theories rest. Together, theology and philosophical ecology invite us to construct a new moral language for our relationship with the Earth — one based on reverence for silence, participation in natural cycles, and fidelity to responsibility.

Convergence Points: The Shared Ground between Religious Teachings and Ethical Theories

Dimension	Religious Teaching	Ethical Theory
Value of Nature	Divine creation, sacred grace, holiness	Intrinsic value, inherent rights
Role of Humanity	Steward, protector, caretaker	Responsible moral agent
Moral Law	Divine commandments, *sharī'ah*, *dharma*	Universal ethical principles
Outcome	Divine reward or punishment, karma	Ecosystem balance, social justice
Principle of Harmony	Humility, gratitude, prayer	Virtue, responsibility, balance

Technical and Ecological Dimension: The Integrity of Ecosystems and the Mechanism of Balance

The Principle of Ecological Balance

Ecologically, nature is a self-sustaining system — a complex network of ecosystems that function through the interaction between biotic and abiotic components (Odum & Barrett, 2005). The silence of nature does not signify stillness but rather reflects a dynamic equilibrium.

Through feedback loops, carrying capacities, and adaptive mechanisms, nature and its resources continuously renew themselves. On the geological timescale, nature's "silence" can be understood as the slowness of grand transformations — processes that unfold over thousands or even millions of years. Thus, nature's silence represents not immobility but a rhythm deeper than human time — a temporal wisdom beyond our own perception.

Human Intervention and the Disturbance of Silence

Since the Industrial Revolution, human interventions in nature and natural resources — including climate change, biodiversity loss, habitat destruction, and pollution — have threatened this silent wholeness. Technological development, while expanding human capability, has often amplified ecological disruption. Contemporary assessments by global organizations (IPBES, 2019; IPCC, 2023) confirm that these threats have intensified, undermining both natural and social resilience.

Humanity's technological voice has grown louder, while nature's silent language of balance has been increasingly drowned out. The result is not only environmental degradation but also an existential estrangement: humankind has silenced the very teacher that once sustained its wisdom.

Listening to Nature's Silence

"Nature is silent because it does not need to shout; it speaks through what it gives."

Within that silence lie generosity, life, balance, and wisdom.

To reconnect with nature and its resources is not to speak to it, but to learn to hear it. Listening becomes an ethical act — a posture of respect, patience, and humility. Through attentive silence, we rediscover that life itself unfolds in cycles of giving and returning.

Chapter Conclusion

Nature, across both religious systems and ethical theories, is regarded as sacred, valuable, and worthy of protection. These approaches converge on a single premise: humanity must not dominate nature but live in harmony with it. The shared message among faiths and philosophies is clear — nature is a trust (amānah), and humans are its guardians.

Consequently, the harm inflicted upon nature and natural resources is not only an environmental violation but also a moral, theological, and existential deviation. Recognizing this reality calls for deeper reflection and transformation at both individual and collective levels.

To restore balance, humanity must return to the principles that once unified spirit, reason, and existence — trust, humility, compassion, gratitude, and justice. Only by realigning with these moral foundations can we renew the bond between humankind and the Earth.

Nature's silence, once again, must be heard — not as absence, but as wisdom; not as stillness, but as balance; not as emptiness, but as life itself.

Chapter References

Akyol, M. (2021). *Doğa Felsefesi ve Hümanizm* [Philosophy of Nature and Humanism]. Retrieved from https://www.researchgate.net/publication/356666451_Doga_Felsefesi_ve_Humanizm

Alkayış, A. (2020). *Çevre ve Etik İlişkisinin Eğitim Felsefesi Bakımından Sorunsallaştırılması* [Problematizing the Relationship Between Environment and Ethics from the Perspective of Educational Philosophy]. Retrieved from https://www.researchgate.net/publication/348357221_Cevre_ve_Etik_Iliskisinin_Egitim_Felsefesi_Bakimindan_Sorunsallastirilmasi

Ardoğan, R. (2012). *İslam'da Çevre Teolojisinin Pratiğe Yansıması: Çevre Ahlakı* [The Reflection of Environmental Theology in Islam into Practice: Environmental Ethics]. *Birey ve Toplum Dergisi, 2*(3), 115–143.

Attfield, R. (1991). *The ethics of environmental concern* (2nd ed.). University of Georgia Press.

Batchelor, S. (1994). *The awakening of the West: The encounter of Buddhism and Western culture*. Riverhead Books.

Bay, S. (2025). *Ramayana ve İnsan-Doğa İlişkisi* [Ramayana and the Human-Nature Relationship]. In *1st International Sarajevo Scientific Research and Innovation Congress*, Bosnia-Herzegovina. Retrieved from https://www.researchgate.net/publication/393553324_RAMAYANA_VE_INSAN-DOGA_ILISKISI

Bernstein, E. F. (2005). The sabbatical year: Environmental wisdom in biblical tradition. In T. Hessel & R. R. Ruether (Eds.), *Christianity and ecology* (pp. 58–74). Harvard University Press.

Berry, T. (1999). *The great work: Our way into the future*. Bell Tower.

Buçukoğlu, S. M. (2020). *Doğa, İnsan ve Sanat Birlikteliği* [The Unity of Nature, Human, and Art]. *Social Sciences Studies Journal, 6*(61), 1888–1892. https://doi.org/10.26449/sssj.2297

Chapple, C. K. (1998). *Hinduism and ecology: The intersection of earth, sky, and water*. Harvard University Press.

Cobb, J. B., & Griffin, D. R. (1976). *Process theology: An introductory exposition*. Westminster Press.

Daly, H., & Farley, J. (2011). *Ecological economics: Principles and applications*. Island Press.

Daily, G. C. (Ed.). (1997). *Nature's services: Societal dependence on natural ecosystems*. Island Press.

De Silva, P. (1998). Environmental philosophy and ethics in Buddhism. *Journal of Buddhist Ethics*. https://doi.org/10.1007/978-1-349-26772-9

Deloria, V. (1994). *God is red: A Native view of religion*. Fulcrum Publishing.

Denny, F. M. (1998). Islam and ecology: A bestowed trust holding the world in trust. *Earth Ethics, 10*(1), 10–11.

Descartes, R. (1641). *Meditations on first philosophy*. Retrieved from https://personal.lse.ac.uk/ROBERT49/teaching/ph103/pdf/Descartes_1641Meditations.pdf

Dindaroğlu, T. (2014). *İnsan ve Doğa İlişkisine "Paraekolojik" Bakış* [A "Paraecological" Perspective on the Human-Nature Relationship]. In *II. Ulusal Akdeniz Orman ve Çevre Sempozyumu*. Retrieved from https://www.researchgate.net/publication/283459581_Insan_ve_Doga_Iliskisine_Paraekolojik_Bakis

Diyanet İşleri Başkanlığı. (2025). *Kur'an Yolu Tefsiri* [Exegesis of the Qur'an]. Retrieved from https://kuran.diyanet.gov.tr/tefsir/

Dönmez, A. (2018). *Biyoetik; Türkiye Bağlamında Doğa, Yaşam, Biyoloji ve İnsan* [Bioethics: Nature, Life, Biology, and Human in the Context of Turkey]. In *24th National Biology Congress*, Manisa, Turkey. Retrieved from https://www.researchgate.net/publication/327673407_Biyoetik_Turkiye_Bagla minda_Doga_Yasam_Biyoloji_ve_Insan

Eliade, M. (1964). *Shamanism: Archaic techniques of ecstasy.* Princeton University Press.

Gündoğan, S. (2022). *Din Felsefesi Açısından Doğa ile Bütünleşmek: Doğa ve Ekolojik Bilincin Oluşumuna Dair Bir Deneme* [Unifying with Nature from the Perspective of Philosophy of Religion: An Essay on the Formation of Ecological Consciousness]. *Ondokuz Mayıs Üniversitesi İlahiyat Fakültesi Dergisi.* https://doi.org/10.17120/omuifd.1104469

Heidegger, M. (1954). *The question concerning technology.* Harper Perennial.

IPBES. (2019). *Global assessment report on biodiversity and ecosystem services.* Intergovernmental Science-Policy Platform on Biodiversity and Ecosystem Services.

IPCC. (2023). *Sixth Assessment Report (AR6).* Intergovernmental Panel on Climate Change. Retrieved from https://www.ipcc.ch/report/sixth-assessment-report-cycle/

Kartal, H. B., & Kartal, A. N. (2020). *İnsan ve Doğa İlişkisi Üzerine Bir Deneme: 1960 ve 1970'lerde Çevrecilik Yaklaşımları* [An Essay on the Human-Nature Relationship: Environmentalism in the 1960s and 1970s]. *Kent Akademisi, 13*(4), 761–770. https://doi.org/10.35674/kent.795143

Kasulis, T. P. (2004). *Shinto: The way home.* University of Hawaii Press.

Kaya, M. (2025). *D'Holbach'ın Felsefesinde Doğa ve İnsan* [Nature and Human in D'Holbach's Philosophy]. *Temaşa: Erciyes Üniversitesi Felsefe Bölümü Dergisi.* https://doi.org/10.55256/temasa.1620716

Lao Tzu. (6th century BCE). *Tao Te Ching* (U. K. Le Guin, Trans.). Metis Yayıncılık.

Leopold, A. (1949). *A sand county almanac and sketches here and there.* Oxford University Press.

McFague, S. (2008). *A new climate for theology: God, the world, and global warming.* Fortress Press.

Miller, J. (2011). *Daoism and nature: The interaction of inner and outer worlds.* Routledge.

Naess, A. (1989). *Ecology, community and lifestyle: Outline of an ecosophy.* Cambridge University Press.

Nasr, S. H. (1996). *Religion and the order of nature.* Oxford University Press.

Norton, B. G. (2005). *Sustainability: A philosophy of adaptive ecosystem management.* University of Chicago Press.

Odum, E. P., & Barrett, G. W. (2005). *Fundamentals of ecology* (5th ed.). Brooks/Cole.

Pepper, D. (1984). *The roots of modern environmentalism.* Routledge.

Pope Francis. (2015). *Laudato Si': On care for our common home.* Vatican Press.

Schwartz, R. H. (2002). *Judaism and global survival.* Lantern Books.

Spinoza, B. (1677). *Ethics.* Retrieved from https://archive.org/details/ethics0000spin_w3j0

Sponsel, L. E. (2012). *Spiritual ecology: A quiet revolution.* Praeger.

Stern, N. (2006). *The economics of climate change: The Stern review.* Cambridge University Press.

Taylor, B. (2010). *Dark green religion: Nature spirituality and the planetary future.* University of California Press. https://doi.org/10.1525/9780520944459

TEEB. (2010). *The economics of ecosystems and biodiversity: Mainstreaming the economics of nature.* Retrieved from https://teebweb.org/

White, L. (1967). The historical roots of our ecologic crisis. *Science, 155*(3767), 1203–1207.

Yeşilçayır, C. (2022). *Viktor E. Frankl'ın Yaşam Felsefesi ve Anlam* [Viktor E. Frankl's Philosophy of Life and Meaning]. *Temaşa: Erciyes Üniversitesi Felsefe Bölümü Dergisi.* https://doi.org/10.55256/temasa.1159089

2nd CHAPTER

HUMAN: REASON, WILL, AND FRAGILITY

Chapter Summary

This chapter analyzes the historical, mental, cultural, and systemic relationship between humans and nature through a multidimensional lens. Although human beings are thinking, planning, and transformative entities, they have also become consumers, forgetful actors, and destroyers—gradually alienating themselves from nature. This alienation has not emerged merely from individual choices, but has been shaped by cultural, economic, political, and psychological structures.

The human–nature relationship is examined in three phases—harmony, dominion, and crisis—emphasizing that today's ecological crises stem from the disruption of this relationship. Multiple factors—such as anthropocentrism, market externalities, insufficient environmental awareness, the short-term nature of political cycles, and the lack of representation of nature and natural resources within existing systems—are discussed in terms of how they collectively contribute to environmental degradation.

Philosophical approaches, from Heidegger to Arne Naess, underline the importance of restoring meaning and responsibility toward nature. Ultimately, the survival of humanity depends on establishing a harmonious, just, and ethical connection with the natural world.

***Keywords*:** Human–nature relationship, Anthropocentrism, Ecological crisis, Collective responsibility, Systemic causes, Harmony with nature

Human Beings: The Maker and the Destroyer; the Mind that Produces, the Hand that Consumes

Human beings, as biological entities endowed with reason, will, and infinite desires, are intrinsically part of nature. Yet, throughout history, they have often perceived themselves as detached from, superior to, or dominant over nature due to their mental and cultural capacities. Humanity has thought, planned, and transformed—but has also consumed, destroyed, and forgotten.

The human relationship with nature and natural resources can be observed in three fundamental stages:

The Period of Harmony: During the hunter-gatherer era, humans lived according to nature's rhythm. This period was largely balanced and sufficient for both sides.

The Period of Dominion: With the Agricultural and Industrial Revolutions and subsequent technological progress, humanity began to control and dominate nature—unfortunately succeeding to a great extent. In this age of dominion, humans blended power with greed, consuming nature and natural resources as if they were limitless commodities meant to satisfy insatiable desires. During this process, nature was transformed into an object of human technical prowess and economic ambition. The essential balance of existence was shaken by an unquenchable desire to possess. Nature ceased to be a shared living space and was reduced to a mere collection of tools reflecting human desires.

The Period of Crisis and Awareness: The long phase of one-sided and boundless exploitation led to severe problems that have now become existential threats to humanity. Contemporary issues such as the climate crisis, depletion of resources, and desertification are direct consequences of humanity's disharmony with nature.

In this period, humanity has divided into two camps: on one side, those who persist in greed and consumption as a mode of existence; on the other, those who recognize the threat and act with an awareness of environmental protection. This second group bears not only the responsibility of repairing a damaged natural world but also the intellectual and moral duty of resisting the culture of excessive consumption.

This duality reflects the inner conflict of modern humans: the tension between the desire to possess and the consciousness of being has led to the erosion of both nature and the human spirit. The struggle today is not merely about reconstructing nature—it is equally about rediscovering humanity's inner measure and existential balance.

Human and Nature Relationship: Competition or Coexistence?

Human beings and nature are not rivals but interdependent and complementary elements. Every human lives within a natural environment composed of living and non-living components and sustained by natural resources. While carrying out all forms of activity, humans utilize and benefit from nature and its resources—yet their actions also affect nature to varying degrees.

However, this interdependence must be based not on power relations, but on balance and sustainability. If humans regard nature merely as a tool for exploitation, and nature in turn perceives humans solely as a threat, such a relationship will yield not harmony but collapse and catastrophe.

When in Harmony:

- Ecosystem services remain sustainable (clean water, fertile soil, clean air, etc.).
- Human health is positively influenced—an effect often referred to as the biophilic effect, meaning the instinctive human inclination to connect with other living beings and nature, which contributes to psychological balance and well-being.
- Societies become more resilient to climate change.
- Cultures draw inspiration from nature, leading to the flourishing of art, ethics, and intellectual capacity within societies.

When in Disharmony:

- Environmental crises emerge—climate change, global warming, drought, desertification, land degradation, sand and dust storms, and so forth.
- The frequency and intensity of natural disasters increase—droughts, floods, hailstorms, frosts, wildfires, storms, etc.
- Food security and safety decline, while poverty, hunger, and migration rise.
- Disputes, conflicts, and wars become more likely, since natural resources are not infinite—they are continuously depleting in ways that affect the welfare of individuals, societies, and nations.
- The human spirit, as it grows distant from nature, becomes increasingly isolated; the search for meaning deepens—a condition often described as the modern existential void.

When harmony offers the possibility of mutual gain, why does humanity persist in choosing loss through discord?

What, then, are the roots of the tension between humans and nature?

The Tension between Nature and Humanity: Short Case Examples

Ganges River (India): The Ganges River, regarded as sacred by Hindus, hosts thousands of religious ceremonies every day. Yet the same river suffers from severe pollution caused by waste, sewage, and industrial chemicals. This paradox between the "sacred" and the "polluted" vividly reveals the contradictions within the superstition–nature relationship.

Amazon Rainforest (South America): Known as the "lungs of the Earth," the Amazon is vital not only for its biodiversity but also as the homeland of numerous Indigenous communities. However, illegal deforestation, mining, and wildfires result in the loss of millions of hectares of forest each year.

Aral Sea (Central Asia): Once the world's fourth-largest lake, the Aral Sea has almost completely dried up due to excessive, misguided, and unplanned agricultural irrigation projects. This example stands in the literature as a tragic symbol of human–nature imbalance—an environmental catastrophe.

Why Do Humans Mistreat Nature?

Throughout human history, nature and natural resources have been perceived both as the source of life and as a threat that must be controlled. In "modern" times, however, this relationship has evolved into a profound crisis—manifested in deforestation, global warming, water scarcity, biodiversity loss, desertification, land degradation, and climate change. But why? Why do humans treat nature and its resources so poorly?

Psychological and cognitive factors

To answer this, it is appropriate to begin with *psychological and cognitive factors.* Among the primary reasons are mental distance and abstraction. As research shows, the human brain struggles to perceive abstract and long-term threats (Weber, 2006). Consequently, issues like global warming or the climate crisis are not often felt as urgent. "When the damage we cause to nature occurs in areas we cannot physically see or sense, our brain does not register it as a threat" (Weber, 2006).

Anthropocentrism, Narcissistic Individualism, and the Culture of Consumption

From the perspectives of anthropocentrism (human-centeredness), narcissistic individualism (self-centered individuals who view themselves as the center of the world), and consumer culture, the "modern" individual always perceives themselves at the center of everything. In modern Western thought, nature and natural resources have been seen as instruments at humanity's disposal. This approach—stretching from Descartes to Locke—has profoundly devalued nature (White, 1967).

"The anthropocentric worldview reduces nature and resources to mere objects existing solely to serve human needs." (White, 1967).

Through materialist paradigms, social media and advertising constantly convey the message that "you are what you consume, and to exist, you must consume more." This narrative strengthens tendencies to commodify, instrumentalize, and exploit nature and natural resources (Kasser, 2002).

Meanwhile, the rapid expansion of urbanization and digitalization has almost entirely severed the human's physical connection with nature, leading to widespread alienation and an inability to understand the natural world (Louv, 2005).

Another critical element to consider is the weakness of cultural narratives, coupled with deficiencies in education, awareness, and ecological literacy. For a long time, modern culture has glorified not harmony and respect for nature, but the conquest of it. In popular culture, nature is frequently portrayed as something "dangerous" or "to be conquered and controlled." Many individuals remain unaware of how ecosystems function, how biodiversity sustains life, or how ecosystem cycles operate. Few realize that a plant or animal species disappearing today might hold the potential for a medicine or cure that could change human life in the future. This lack of ecological awareness often leads to irresponsible human behavior (Orr, 1992).

Economic and Systemic Reasons

Among the economic and systemic factors, global inequalities, injustices, and the capitalist production–consumption cycle stand out. In the context of development and inequality, less-developed and developing countries are often more dependent on nature, the environment, and natural resources. However, global capital exploits these resources while creating political, social, economic, and environmental injustices within those same nations (Martinez-Alier, 2002).

Among all creatures, humans are both the most honorable and the most greedy, ambitious, and insatiable. Humanity is the only species that desires far more than what it needs to survive. Humans utilize nature and natural resources in their production of goods and services, and afterward, they discharge their wastes back into the environment. Thus, nature becomes both exploited and polluted (Başol et al., 2007).

Consequently, for the sake of capital accumulation, nature and its resources are continually used as both a raw material reservoir and a waste disposal site. Profit maximization in commercial activities rarely takes environmental destruction into account (Foster, 2000).

Externalities and the Problem of Pricing

Another critical issue is externalities and pricing failures. Capitalist and materialist systems often generate negative externalities. The environmental damage inflicted upon nature and resources in the pursuit of profit is typically paid for by others.

“The burden of the global environmental crisis is placed on the shoulders of those least responsible.” (IPCC, 2022).

Capitalist market systems generally fail to "price in" the damages done to nature. For instance, cutting down a forest generates commercial profit, yet the forest's carbon sequestration value remains outside the realm of market pricing (Stern, 2006).

"The cost of destroying natural capital is, in most cases, systematically ignored." (Stern, 2006).

This problem is also reinforced by social and cultural norms that regard nature and its resources as infinite, boundless, and free. Until recently, cultural beliefs portrayed nature as self-renewing and inexhaustible. Such misconceptions have legitimized unsustainable patterns of individual consumption.

From the perspective of collective responsibility, Garrett Hardin's (1968) Tragedy of the Commons illustrates that shared resources tend to be overused because no individual feels personally accountable for their preservation.

Political and Institutional Reasons

Among political and institutional causes, short-term political cycles are the most prominent. The question of why politics and politicians do not give sufficient importance to nature and natural resources has multifaceted historical, sociopolitical, temporal, economic, and ideological dimensions.

Politics generally concerns itself with the visible and short-term—projects that produce tangible results within electoral periods. Since politicians operate according to election cycles of roughly four to five years, they tend to focus on policies that yield immediate and visible benefits to voters.

In contrast, environmental and resource-based policies typically produce long-term outcomes, often invisible within a single political term. As a result, policymakers frequently neglect nature-related investments because they are under pressure to demonstrate rapid results that will please their electorate. In such cases, politicians and voters may be satisfied, but the losers are almost always the environment, nature, and natural resources.

For example, forest conservation or carbon reduction initiatives are long-term projects whose benefits are not immediately visible, whereas infrastructure projects such as roads, bridges, dams, and canals produce quick, visible results and win votes. Consequently, political leaders often hesitate to invest in environmental sustainability, as such investments provide "invisible" benefits that do not translate into immediate electoral success (Jacobs, 2016). Yet, the protection of nature and natural resources yields enduring and meaningful benefits over time.

Furthermore, many policymakers perceive environmental sustainability as conflicting with economic growth. In developing countries especially, environmentally damaging projects—such as mining, large dams, transport, infrastructure, and industrial investments—are often justified in the name of "development." While such projects may be necessary, they must be undertaken within a rational, equitable, and sustainable framework.

In political rhetoric, environmental costs are often marginalized in favor of economic growth, despite the moral appeal of sustainability (Dryzek & Pickering, 2018). Another contributing factor is the weakness of nature-based knowledge and environmental awareness among politicians. Most political actors maintain limited engagement with the natural sciences; their decisions are instead shaped by ideological, economic, or security-oriented considerations.

As a result, strong, rational bridges between scientists and policymakers rarely exist. One major reason for the failure of environmental policies lies in the fact that decision-makers are often detached from the natural sciences (Meadowcroft, 2002).

Lobbying, Private Sector Pressure, and the Political Representation of Nature

The influence of lobbying and private sector interests requires special attention. Industries that cause significant environmental harm—such as mining, oil, coal, and construction—exert enormous influence over political actors. Driven by commercial motivations, these lobbies systematically obstruct the implementation of nature-friendly policies. Moreover, systems of political financing have become increasingly dependent on such interest groups.

"Industrial lobbies systematically weaken environmental regulations and, by influencing decision-makers, prioritize private interests over the public good." (Brulle, 2014).

The lack of political representation for nature must also be recognized as a profound irony. Unlike humans, the environment, nature, and natural resources cannot vote or lobby. Therefore, within representative democracy, they possess no direct voice. Ecological concerns often remain marginalized within the structural boundaries of political systems.

"Nature is an actor without a voice within systems of representation, and this silence paves the way for environmental destruction." (Latour, 2004).

Another major problem is the weakness of public demand for environmental protection. Citizens and societies often fail to prioritize environmental issues due to factors such as political ideology, welfare levels, income, and education. Where there is no strong public demand for environmental policy, political will rarely emerges.

"The success of environmental policies is directly related to the demands of citizens." (Inglehart, 1995).

Why Does Nature Remain In The Background?

Why do politicians fail to give sufficient importance to nature? Because:

- Short-term electoral calculations often hinder long-term investments in the environment, nature, and natural resources.
- The rhetoric of economic growth has surpassed environmental, ecological, and sustainable resource management priorities in many parts of the world.
- The science–policy bridge has not developed adequately on a rational basis (or has not been allowed to develop for various reasons).
- Private sector and lobbying pressures significantly obstruct public environmental, nature, and natural resource policies.
- Nature, the environment, and natural resources lack direct political representation within current political systems.
- For various reasons, there is insufficient public demand for environmental protection and sustainability.

Therefore, the existing situation is not only a problem of politicians but also a structural problem of the system, society, and individuals.

Another important issue to consider is insufficient environmental legislation and deficiencies in implementation. Globally, regardless of development level, there are numerous laws and practices intended to protect the environment, nature, and natural resources. However, in many countries, environmental protection laws and enforcement mechanisms are either weak, ineffective, or obstructed by pressure from interest groups. Especially in developing countries, monitoring and inspection systems are inadequate (OECD, 2011).

Categorical Distribution of the Reasons

Category	Reason	Explanation
Psychological	*Psychological Distance*	Inability to perceive abstract and long-term threats clearly
Economic	*Market Externalities*	Failure to price the true value of nature
Political	*Short-Term Thinking*	Policy decisions indexed to electoral cycles
Philosophical	*Anthropocentrism*	Devaluation of nature through a human-centered worldview
Cultural	*Alienation from Nature*	Estrangement from nature through digitalization and urbanization
Educational	*Lack of Ecological Awareness*	Insufficient knowledge of nature, ecosystem cycles, and interdependence
Global	*Inequality and Injustice*	Exploitation of natural resources and unequal distribution of environmental burdens

Behind humanity's mistreatment of nature lie not merely ignorance but deeper, systemic, and structural causes. Confronting these causes requires not only individual ethics but also collective responsibility, systemic transformation, and a reconstruction of values.

To Know, To Feel, To Understand, To Respect, To Change...

In the context of a person remembering themselves and of the "Living Earth," nature reminds humanity of its essence: Soil symbolizes creation and mortality, Water represents life and the fluidity of the soul, Air reflects one's perception, existence, and thoughts, Fire embodies human desire — the flame that can either warm the spirit with love or consume it into ashes.

Similar to Martin Heidegger's assertion that "Being attains meaning when it takes root in space and time" (Heidegger, 1977), humans too can truly experience and comprehend their existence and purpose only when they take root within nature itself.

"What Kind Of Transformation Is Possible?" — List of Recommendations

Ecological Literacy Education: Individuals should be enabled to understand natural cycles, biodiversity, and methods of environmental protection.

Nature-Based Learning Models: Schools should promote observation- and experience-based education integrated with nature.

Eco-Friendly Urban Planning: Instead of concrete expansion, cities should prioritize green infrastructure, public transport, bicycle networks, and nature-integrated living spaces.

Slow Living and Sustainable Consumption Models: A lifestyle culture based on needs and sustainability, rather than consumption and excess, should be encouraged.

Nature Ethics and Spiritual Connection: People should be encouraged to form not only physical but also moral and spiritual bonds with nature — through activities such as forest therapy, nature walks, and permaculture (an ecological design philosophy that applies sustainable, ecosystem-based principles to human settlements).

Humanity and Nature are Two Faces of a Shared Existence

Humans can exist not by separating from or destroying nature, but by understanding, acting justly, and establishing a moral relationship with it.

Arne Naess's Deep Ecology perspective argues that humanity is connected to nature not only biologically, but also morally and ethically:

> "We must learn not to live for nature, but with nature."

Thus, the two are not rivals, but complementary realms of existence. Harmony with Nature → sustainable life, happiness, and abundance. Disharmony → crisis, collapse, and isolation.

The choice is yours…

Chapter Conclusion

Humanity and nature are not two separate entities; they are two faces of the same life and the same cycle. The human being who distances themselves from nature risks losing not only ecosystems but also their soul, meaning, and future.

Therefore, transformation must occur not only at the level of individual awareness but also at collective and systemic levels. From politics to economics, from education to culture, every structure must be reconstructed in harmony with nature and natural resources.

Because within a system that conflicts with nature, neither genuine development nor lasting prosperity and happiness are possible.

Chapter References

Altınbaş, M. (2023). *Psikolojik Kırılganlık* [Psychological Vulnerability]. In *Örgütsel Davranışta "Psikolojik" Başlıklı Konular*. Gazi Kitabevi, Ankara. Retrieved from https://www.researchgate.net/publication/376954027_PSIKOLOJIK_KIRILGANLIK

Başol, K., Durman, M., & Önder, H. (2007). *Doğal Kaynakların ve Çevrenin Ekonomik Analizi* [Economic Analysis of Natural Resources and the Environment]. Alfa Aktüel Yayınları, Bursa.

Berkes, F. (2008). *Sacred Ecology*. Routledge, Oxfordshire, UK. https://doi.org/10.4324/9780203928950

Brulle, R. J. (2014). Institutionalizing delay: Foundation funding and the creation of US climate change counter-movement organizations. *Climatic Change, 122*(4), 681–694. https://doi.org/10.1007/s10584-013-1018-7

Capra, F. (1996). *The Web of Life: A New Scientific Understanding of Living Systems*. Anchor Books, Harpswell, ME, Canada.

https://books.google.com.tr/books/about/The_Web_of_Life.html?id=idDh0VYhsCIC

Çakmak, M. (2021). Skolastik Felsefede Ahlâkın İnsan Doğası ve Tanrı ile İlişkisi [The Relationship Between Morality, Human Nature, and God in Scholastic Philosophy]. *Sakarya Üniversitesi İlahiyat Fakültesi Dergisi (SAUIFD), 23*(44). https://doi.org/10.17335/sakaifd.985830

Diş, B. S. (2021). Descartes'a Göre Edilgiler, Akıl ve Ahlak Arasındaki İlişki [According to Descartes: The Relationship Between Passions, Reason, and Morality]. *Felsefe Dünyası Dergisi, 74*, 5–26.

Dryzek, J. S., & Pickering, J. (2018). *The Politics of the Anthropocene*. Oxford University Press, UK. Retrieved from https://global.oup.com/academic/product/the-politics-of-the-anthropocene-9780198809623

Ekiz, S. (2022). İnsan haklarının evrenselliği ve kültürel görecelik [Universality of Human Rights and Cultural Relativism]. Retrieved from https://www.researchgate.net/publication/388109277_insan_haklarinin_evrenselligi_ve_kulturel_gorecelik

Folke, C., Colding, J., & Berkes, F. (2003). Building resilience and adaptive capacity in social-ecological systems. In *Navigating Social-Ecological Systems* (pp. 352–387). Cambridge University Press, UK.

Foster, J. B. (2000). *Marx's Ecology: Materialism and Nature*. Monthly Review Press. https://doi.org/10.2307/3341498

Genç, C. (2025). İlk Dönem Kelam-Felsefe İlişkisinde İman Akıl ve İrade [Faith, Reason, and Will in Early Kalam-Philosophy Relations]. *Türkiye İlahiyat Araştırmaları Dergisi, 9*(3), 794–817. https://doi.org/10.32711/tiad.1577871

Hardin, G. (1968). The tragedy of the commons. *Science, 162*(3859), 1243–1248.

Heidegger, M. (1977). *The Question Concerning Technology and Other Essays*. Harper & Row. Retrieved from https://www.amazon.com/QuestionConcerningTechnologyOtherEssays/dp/0061319694

Inglehart, R. (1995). Public support for environmental protection: Objective problems and subjective values in 43 societies. *PS: Political Science & Politics, 28*(1), 57–72.

IPCC. (2022). *Sixth Assessment Report (AR6).* Intergovernmental Panel on Climate Change. Retrieved from https://www.ipcc.ch/assessment-report/ar6/

Jacobs, A. M. (2016). Policy making for the long term in advanced democracies. *Annual Review of Political Science, 19*, 433–454.

Kasser, T. (2002). *The High Price of Materialism.* MIT Press, US. Retrieved from https://mitpress.mit.edu/9780262611978/the-high-price-of-materialism/

Kılıç, E. (2022). Değerin Doğasına Dair Üç Mesele: Güç, İrade ve Eylem [Three Issues on the Nature of Value: Power, Will, and Action]. *Eskiyeni*, https://doi.org/10.37697/eskiyeni.1033394

Latour, B. (2004). *Politics of Nature: How to Bring the Sciences into Democracy.* Harvard University Press. Retrieved from https://www.amazon.com.tr/PoliticsNatureBringSciencesDemocracy/dp/0674013476

Leopold, A. (1949). *A Sand County Almanac.* Oxford University Press, UK. Retrieved from https://global.oup.com/academic/product/a-sand-countyalmanac9780197500262

Louv, R. (2005). *Last Child in the Woods: Saving Our Children from Nature-Deficit Disorder.* Algonquin Books, New York, US. Retrieved from https://www.environmentandsociety.org/mml/last-child-woods-saving-our-children-nature-deficit-disorder

Martinez-Alier, J. (2002). *The Environmentalism of the Poor.* Edward Elgar Publishing, Massachusetts, US. Retrieved from https://cdn.unrisd.org/assets/library/papers/pdf-files/martinezalier.pdf

Meadowcroft, J. (2002). Politics and scale: Some implications for environmental governance. *Landscape and Urban Planning, 61*(2–4), 169–179. https://doi.org/10.1016/S0169-2046(02)00111-1

Naess, A. (1973). The shallow and the deep, long-range ecology movement. *Inquiry, 16*(1–4), 95–100.

OECD. (2011). *Environmental Outlook to 2050: The Consequences of Inaction.* OECD Publishing. https://doi.org/10.1787/9789264122246-en

Orr, D. W. (1992). *Ecological Literacy: Education and the Transition to a Postmodern World.* SUNY Press, Albany, New York, US.

Stern, N. (2006). *The Economics of Climate Change: The Stern Review.* Cambridge University Press, UK.

Weber, E. U. (2006). Experience-based and description-based perceptions of long-term risk: Why global warming does not scare us (yet). *Climatic Change, 77*(1–2), 103–120.

White, L. (1967). The historical roots of our ecological crisis. *Science, 155*(3767), 1203–1207.

Yazgünoğlu, K. C. (2025). İnsan ve Ötesi: Posthümanist Düşüncede İlişkisel Öznellik ve Dolanık Ontolojiler [Human and Beyond: Relational Subjectivity and Entangled Ontologies in Posthumanist Thought]. *FLSF Felsefe ve Sosyal Bilimler Dergisi.* https://doi.org/10.53844/flsf.1728458

Yazıcı, Ş. (2024). *Tanrı, İyilik ve İrade* [God, Goodness, and Will]. Küre Yayınları. Retrieved from https://www.researchgate.net/publication/386484850_Tanri_Iyilik_ve_Irade

3rd CHAPTER

THE REBELLION OF NATURE: THE MULTIPLICITY OF CRISES

"The danger for humanity is not only the change of nature, but the forgetting of its responsibility toward it." – Arne Naess

Chapter Summary

This chapter analyzes the pressure of human activities on the environment, nature, and natural resources, revealing the multilayered structure of the global climate crisis. The use of fossil fuels, deforestation, intensive agriculture and livestock production, uncontrolled consumption habits, and greenhouse gas emissions from transportation systems have disrupted the Earth's energy balance, rapidly heating the planet.

Climate change is not limited to rising temperatures; it also triggers multifaceted crises such as the increasing frequency and intensity of extreme weather events, glacial melting, sea level rise, and decreases in agricultural productivity, food insecurity, mass migration movements, and security risks.

By supporting these cause-and-effect relationships with numerical data, scientific analyses, and global examples, this chapter provides a holistic perspective. Framed by the expression "Nature is not merely being harmed; it is responding," the chapter emphasizes that nature has broken its silence — it now acts as an active subject, compelling humanity to take responsibility.

Ultimately, this section stands both as a warning sign for humanity and a call for collective responsibility toward a sustainable future.

Keywords**:** Global Warming, Climate Change, Greenhouse Gases, Consumption Culture, Extreme Weather Events, Climate Migration

The Anatomy of Ecological Crises: What Has Happened, What Is Happening?

"Nature is not only being harmed — it is responding."

The 21st century is recorded as an era in which humanity's destruction of nature and natural resources has surpassed all limits — and nature has ceased to remain silent. Today, nature and resources are not merely passive entities being destroyed; they are active subjects, capable of responding and feeding back in their own language.

The global phenomena of our time — disruptions in climate systems, extreme weather events, agricultural and food crises, migration movements, and emerging security risks — expose, in a comprehensive way, the necessity of recognizing reality and taking urgent action.

Quantitative data and scientific evidence reveal how clearly the planet communicates its distress — yet how indifferent humanity remains to this message. These ongoing crises challenge the anthropocentric worldview and underscore the necessity of building a new, nature-aligned mode of existence.

Global Warming and Climate Change: The Silent Fire of Human History

Global warming refers to the long-term rise in Earth's surface temperatures resulting from increased concentrations of greenhouse gases in the atmosphere. Climate change, in contrast, encompasses not only temperature increases but also changes in precipitation regimes, the frequency and intensity of extreme weather events, droughts, and rising sea levels — thus representing a broader set of transformations.

In both processes, humanity has disrupted the planet's natural climatic cycles through uncontrolled and excessive activity — in ways that may be nearly irreversible.

Causes: Humanity's Anomalous Interaction with the Atmosphere

Excessive Use of Fossil Fuels (Coal, Oil, and Natural Gas): Since the Industrial Revolution, the use of coal, oil, and natural gas as primary energy sources has steadily increased. This has led to the continuous release of large quantities of greenhouse gases, primarily carbon dioxide (CO_2), into the atmosphere.

Greenhouse gases are compounds that trap heat by preventing solar radiation reflected from the Earth's surface from escaping back into space. These include carbon dioxide (CO_2), methane (CH_4), nitrous oxide (N_2O), hydrofluorocarbons (HFCs), perfluorocarbons (PFCs), and sulfur hexafluoride (SF_6) — all possessing strong heat-trapping properties.

Although water vapor is the most abundant greenhouse gas in the atmosphere, it is not the initiator of warming. Rather, it acts as a positive feedback mechanism: following the initial warming caused by CO_2 and CH_4, humidity increases by approximately 7% per °C, further amplifying the warming. Thus, policy efforts to control climate change prioritize reducing CO_2 and CH_4 emissions rather than directly regulating water vapor.

According to the Intergovernmental Panel on Climate Change (IPCC), most anthropogenic warming originates from CO_2 and CH_4. Roughly two-thirds of global warming can be attributed to CO_2, while one-third stems from methane. Aerosols, meanwhile, exert a partial cooling effect, masking some of the warming.

Sectoral Distribution of Greenhouse Gas Emissions: Analysis of global emissions reveals that the energy sector is the dominant source. Electricity and heat production account for about 30% of total emissions, followed by transportation (14%) and manufacturing and construction (13%). Direct fuel use in buildings contributes roughly 7%, while agriculture (12%) and waste management (3%) represent the main non-energy sources. Today, over 75% of human-induced CO_2 emissions are directly linked to fossil fuel combustion (IEA, 2021).

Carbon Dioxide and the Carbon Cycle: Carbon dioxide is a fundamental component of the carbon cycle. It is released into the atmosphere through the oxidation or combustion of carbon-containing substances. Plants, in turn, absorb CO_2 during photosynthesis and convert it into cellulose and other carbohydrates. Through food chains, carbon circulates among living organisms.

When organisms die, their remains accumulate underground under anoxic conditions, gradually forming fossil fuels. Meanwhile, CO_2 dissolved in rainwater contributes to the formation of carbonate rocks, such as limestone. When limestone is heated, CO_2 is released into the atmosphere but is later reabsorbed during the hardening of lime — completing a natural cycle.

Under natural conditions, this carbon cycle maintained atmospheric CO_2 levels in equilibrium for centuries. However, the Industrial Revolution disrupted this balance. Intensive fossil fuel use for energy, transportation, and heating has driven CO_2 emissions to record levels, rapidly increasing atmospheric CO_2 concentrations.

Simultaneously, the loss of forests and green vegetation has further aggravated the problem — both adding extra emissions and reducing terrestrial carbon sinks, thereby weakening the planet's capacity to naturally reabsorb CO_2.

As a result of these dual imbalances, atmospheric CO_2 concentrations — which were about 280 ppm in 1850 — rose to 380–400 ppm by 2000. Alarmingly, this upward trend has continued to accelerate (AURA, 2025). By the end of 2024, the atmospheric CO_2 concentration reached 425 ppm. To limit global warming to below 2°C by the end of this century, atmospheric CO_2 levels must be stabilized between 450–550 ppm. Yet the current annual global average increase in CO_2 concentration is already around 2–3 ppm (Tolunay, 2025).

Deforestation and the Decline of Carbon Sinks: Forests are natural carbon sinks that absorb carbon dioxide from the atmosphere through photosynthesis, storing it in their biomass and soils. However, deforestation caused by agriculture, livestock production, and urbanization, along with forest fires, has drastically reduced this capacity and simultaneously triggered new carbon emissions into the atmosphere (FAO, 2020). Deforestation, which amplifies the impacts of climate change, continues to pose a major global threat. Since 1990, approximately 178 million hectares of forest have been lost worldwide. Today, deforestation and forest degradation threaten this natural balance and accelerate global warming. The destruction of forests represents a dual threat to nature: first, it releases the stored carbon into the atmosphere; second, it reduces the potential for future carbon sequestration (Pan et al., 2011).

Globally, forests absorb about 2.4 billion tons of carbon dioxide annually (FAO, 2020). However, deforestation and land degradation release around 3.3 billion tons of CO_2 into the atmosphere each year (IPCC, 2021). Between 1990 and 2020, the world lost approximately 178 million hectares of forest—roughly twice the land area of Türkiye (FAO, 2020). For instance, in the Amazon Rainforest, it was estimated that between 2001 and 2019, 60 billion tons of carbon were absorbed from the atmosphere, but during the same period, 13.2 billion tons of carbon were released due to deforestation (Gatti et al., 2021). In 2022 alone, 4.1 million hectares of tropical forests were lost, resulting in emissions equivalent to 2.7 billion tons of carbon dioxide (Global Forest Watch, 2023).

In Türkiye, although forest cover appears to have increased from 22.3 million hectares in 2000 to 23.36 million hectares in 2025 (TÜİK, 2025), the functional quality of these areas as efficient carbon sinks has declined. Productive, mature forests are being replaced by fragmented and younger structures (OGM, 2022). During the 2021 wildfires in Türkiye, approximately 140,000 hectares of forest were burned, leading to an estimated 8–10 million tons of carbon dioxide emissions (TEMA, 2021).

The loss of forests affects not only the carbon cycle but also several climate-regulating systems such as the albedo effect (the ability of surfaces to reflect sunlight), the moisture cycle, and local temperature regulation (Bonan, 2008). Reduced forest cover leads to less CO_2 absorption while intensifying drought conditions and urban heat island effects on local scales. Particularly in tropical regions, deforestation disrupts regional climate stability and accelerates global warming (Malhi et al., 2008).

Methane Emissions from Agriculture and Livestock: The agricultural and livestock sectors are responsible for approximately 40% of global methane emissions. Methane, one of the primary gases directly linked to these sectors, contributes significantly to climate change through industrial agriculture and intensive livestock production. Although methane remains in the atmosphere for a shorter period than carbon dioxide, its Global Warming Potential (GWP) is 25 to 84 times higher than that of CO_2 (IPCC, 2021; IPCC, 2022). Thus, methane emissions exert a disproportionately large impact on global warming.

As of 2021, global methane emissions were estimated at 570 million tons per year. Of this total, approximately 40% (227 million tons) originated from agriculture, 35% from fossil fuels, 20% from waste management, and 5% from other sectors (UNEP & CCAC, 2021).

The two largest agricultural sources of methane are:

Ruminant livestock digestion: The global livestock sector accounts for around 14.5% of total greenhouse gas emissions, with 27% of this share attributed to methane (Gerber et al., 2013). On average, a single cow emits 70–120 kg of methane per year (Gerber et al., 2013). At a global scale, enteric fermentation from livestock produced approximately 102 million tons of methane annually as of 2020 (FAO, 2020).

Anaerobic conditions in rice paddies: Rice fields cultivated using traditional flooding methods create anaerobic environments that produce methane. 10–12% of agricultural methane emissions come from this source (Saunois et al., 2020). About 90% of global rice production occurs in Asia, where rice cultivation emits 50–60 million tons of methane per year (Yan et al., 2009). These emissions can be reduced by 30–70% through improved irrigation and intermittent drying techniques (FAO, 2021).

In Türkiye, as of 2021, the agricultural sector accounted for 13.5% of total greenhouse gas emissions, with a significant portion consisting of methane and nitrous oxide from agricultural activities (TÜİK, 2022). Methane emissions from livestock production constitute about 45% of total agricultural emissions, with Konya, Şanlıurfa, and Erzurum being among the provinces with the highest methane emissions (Tarım Orman Strateji, 2020).

The Impact of the Culture of Consumption and Transportation on Global Warming: Urbanization and the increasing use of private vehicles have significantly raised greenhouse gas emissions, particularly in the transportation sector. This demonstrates the dependency of transportation and energy systems on carbon-based sources. Moreover, consumption-oriented lifestyles continuously increase energy demand and, consequently, greenhouse gas emissions. Global warming is fueled not only by production systems but also by consumption habits.

The culture of consumption promotes the overuse of natural resources, single-use products, fast fashion, food waste, and high energy consumption — all of which intensify greenhouse gas emissions. Transportation, in this context, serves both as a consequence of and a major contributor to this system. Growing mobility, private car dependence, air travel, and e-commerce logistics have dramatically expanded the global carbon footprint (Jackson, 2009; Gössling et al., 2020).

The Impact of Consumption Culture on Emissions: Approximately 60% of global greenhouse gas emissions are driven by consumer demand (Ivanova et al., 2016). The per capita carbon footprint in high-income countries is 10 to 50 times larger than in low-income nations (Chakravarty et al., 2009). According to a 2015 study, household consumption alone contributes around 72% of total global greenhouse gas emissions (Ivanova et al., 2016). The fast fashion industry produces more carbon emissions than international aviation and shipping combined — around 1.2 billion tons of CO_2-equivalent annually (Ellen MacArthur Foundation, 2017). Globally, food consumption ranks first among household emission sources, followed by housing (heating and electricity), transportation, and clothing.

Transportation Sector and Global Warming: The transportation sector accounts for approximately 24% of global greenhouse gas emissions. Of these, 72% stem from road transport, 11% from aviation, and 10% from maritime transport (IEA, 2022). In 2021, transport-related CO_2 emissions reached 7.7 billion tons (IEA, 2022). Private car use produces 4–6 times more emissions per person compared to public transport (Gössling et al., 2020).

Carbon Dioxide Emissions per Kilometer of Travel (Per Person):

- Car (single occupant): 271 g of carbon dioxide
- Bus: 101 g of carbon dioxide
- High-speed train: 41 g of carbon dioxide
- Airplane (economy class): 285 g of carbon dioxide (EEA, 2021).

E-Commerce and the Logistics Chain: With the rise of e-commerce, last-mile deliveries and rapid logistics systems have significantly increased urban emissions. It is estimated that by 2030, carbon dioxide emissions from urban deliveries related to e-commerce will increase by 32% (World Economic Forum, 2020).

Transportation and Consumption Trends in Türkiye

As of 2021, the transportation sector in Türkiye accounted for 15.6% of total greenhouse gas emissions (TÜİK, 2022). In 2000, transportation-related emissions in Türkiye were approximately 42 million tons of CO_2 equivalent, while by 2021, this figure had risen to 85 million tons. This increase is largely influenced by the rapid growth in the number of motor vehicles in urban areas and the insufficient use of public transportation. As of 2023, the total number of registered motor land vehicles in Türkiye reached 27 million (TÜİK, 2023).

The Balance of the Planet Is Being Shaken

Global Temperature Rise: The rise in global temperature is one of the most visible indicators of climate change. Human activities—especially fossil fuel consumption, industrialization, deforestation, and high consumption patterns—have increased the concentration of greenhouse gases in the atmosphere (particularly CO_2, CH_4, and N_2O). This has disrupted the Earth's energy balance, leading to a clear increase in average temperatures (IPCC, 2021).

As of 2023, the planet has warmed by approximately 1.1–1.2°C above pre-industrial levels. This increase is rapidly approaching the critical threshold of 1.5°C (IPCC, 2021; WMO, 2023).

Numerical Facts: How Much Have We Warmed?

- Compared to the period 1880–1900, the global average surface temperature has increased by approximately 1.2°C as of 2023 (NASA, 2024).
- Over the 170-year period between 1851 and 2023, global warming has increased by 1.36°C (Yılmaz, 2025).
- Between 1981 and 2023, global warming increased by 1.14°C (Yılmaz, 2025).
- The year 2023 was the warmest year ever recorded, with global average temperatures 1.48°C above pre-industrial levels (1850–1900) (Copernicus Climate Change Service [C3S], 2024).
- The 2011–2020 decade was 1.1°C warmer than the pre-industrial period and marked the hottest decade on record (IPCC, 2021).
- The Arctic is warming at least twice, and in some periods more than three times, faster than the global average (a phenomenon known as Arctic Amplification), with an average annual increase of 0.3°C (AMAP, 2021).
- If policies implemented by the end of 2020 continue unchanged, global warming is expected to reach an average of 3.2 [2.2–3.5]°C by 2100 (Yılmaz, 2025).
- In Türkiye, the average temperature between 2071–2100 is projected to be 1.9°C higher under the optimistic SSP2-4.5 scenario, and 3.9°C higher under the pessimistic SSP5-8.5 scenario compared to the 1980–2014 period (Yılmaz, 2025).

Geographical Distribution of Temperature Rise

- The Northern Hemisphere, particularly Europe, North America, and the Arctic, has experienced a more pronounced rise in temperature.
- The Mediterranean Basin, including Türkiye, has been defined by the IPCC (Intergovernmental Panel on Climate Change) as one of the climate change "hot spots"—regions most vulnerable to warming. In Türkiye, the average temperature has increased by more than 1.5°C over the past 50 years (MGM, 2022).
- In 2023, the average temperature in some parts of Türkiye was measured to be +2.5°C above the long-term average (MGM, 2024).

Impacts of Rising Temperatures

- The frequency and intensity of droughts are increasing, agricultural productivity is declining, and pressure on water resources—both in quantity and quality—is intensifying (Trenberth et al., 2014; Sevinç et al., 2019; Parlakçı Doğan et al., 2020; Aydoğdu et al., 2021).
- For Türkiye, precipitation projections indicate a decrease in annual average rainfall in the Aegean and Mediterranean regions across all forecast periods, while an increase is expected in the Eastern Black Sea and Eastern Anatolia regions. Consequently, drought risks are rising in the southern half of the country (Yılmaz, 2025).

Extreme weather events—including heatwaves, wildfires, frost events, and floods—are occurring more frequently and becoming more destructive.

- For example, during the summer of 2023, some parts of Europe recorded daytime temperatures exceeding 45°C (WMO, 2023).
- The summer of 2022 was the hottest ever recorded in Europe. Between May 30 and September 4, an estimated 61,672 people (confidence interval: 37,643–86,807) lost their lives due to extreme heat (Ballester et al., 2023).
- The majority of these deaths were not direct heatstroke's, but rather triggered by extreme temperatures worsening pre-existing cardio-respiratory diseases (World Health Organization [WHO], 2024).
- Europe witnessed another heat record in the summer of 2024 (European Centre for Medium-Range Weather Forecasts / Climate Change Service [ECMWF/C3S], 2025).

- Approximately 56% of heat-related deaths during the summer of 2022 were attributed to human-induced climate warming (Beck et al., 2024).
- Without appropriate adaptation and protection measures, the number of heat-related deaths in Europe is projected to reach around 120,000 per year by 2050 (WHO/Europe, 2025; see also Ballester et al., 2023).
- Ocean surface temperatures are also rising, creating significant adverse impacts on marine ecosystems, coral reefs, and fish stocks (IPCC, 2021).

If Global Warming Reaches 2°C Instead of 1.5°C:

- The number of people exposed to extreme heat will increase by an additional 700 million worldwide.
- The loss of coral reefs will rise from 70% to 99% (Hoegh-Guldberg et al., 2018).

Critical Thresholds and Political Targets

- The Paris Agreement (2015) aims to keep the global temperature rise below 2°C, and preferably limited to 1.5°C (UNFCCC, 2015).
- However, if current policies continue, the temperature increase by the end of the century is expected to range between 2.7°C and 3.2°C (Climate Action Tracker, 2023), indicating that existing actions remain far from achieving the Paris Agreement's 1.5°C or 2°C targets.
- To prevent exceeding the 1.5°C threshold, global greenhouse gas emissions must be reduced by 43% by 2030 (IPCC, 2022).

Paris Agreement and Country-Based Commitment Comparisons

The main goal of the Paris Agreement is to limit global temperature rise to below 2°C and preferably to 1.5°C. To achieve this, countries have prepared Nationally Determined Contributions (NDCs). However, these commitments differ significantly in both scope and implementation among nations.

People's Republic of China

- Share of Global Emissions: Approximately 29% (Ranked 1st globally)
- Paris Agreement Participation: 2016
- Net-Zero Target: 2060

Commitments:

- Reach peak emissions before 2030
- Reduce carbon intensity by 60–65% relative to 2005 levels by 2030
- Install 1,200 GW of renewable energy capacity by 2030

Status:

As of 2023, coal-based power generation remains high. However, in terms of renewables, China is the global leader in both solar and wind energy development.

European Union (EU)

- Share of Global Emissions: Approximately 7–8%
- Paris Agreement Participation: 2016
- Net-Zero Target: 2050

Commitments:

- Reduce greenhouse gas emissions by 55% compared to 1990 levels by 2030
- Implement the European Green Deal, driving transformation across energy, agriculture, transportation, and industry sectors

Status:

By 2023, renewables accounted for 45.3% of total electricity consumption and 24.5% of total energy use. The EU continues its rapid coal phase-out and operates a carbon pricing policy through its Emissions Trading System (ETS).

Republic of Türkiye

- Share of Global Emissions: Approximately 1.1%
- Paris Agreement Participation: 2021 (late ratification)
- Net-Zero Target: 2053

Commitments:

- Reduce emissions by 41% relative to the reference scenario by 2030
- Achieve an annual reduction of 17 million tons of CO_2 equivalent by 2030

Status:

Türkiye's "reference scenario" approach implies continued emissions growth rather than absolute reductions. However, following the 2023 Climate Council, several sectoral strategies—particularly on renewable energy, energy efficiency, and transportation—have started to take shape.

The Climate Law, adopted in July 2025, encompasses provisions on greenhouse gas mitigation, climate adaptation, related planning and implementation tools, financial mechanisms, permitting and supervision processes, and the legal and institutional framework governing climate action in Türkiye.

Although the Paris Agreement provides a universal framework, countries' contributions are not legally binding and rely on national will. China's continued dependence on coal, the EU's pioneering role, and Türkiye's emerging emission policies all illustrate that achieving global goals depends on stronger, more transparent, and traceable implementation mechanisms.

Country-Based Comparison

Country	Share of Global Emissions	Net-Zero Target	2030 Target	Implementation Status
China	29%	2060	60–65% reduction in emission intensity	Rapid growth in solar/renewable energy, but continued coal dependency
European Union	7–8%	2050	55% reduction compared to 1990	ETS, Green Deal, coal phase-out
Türkiye	1.1%	2053	41% reduction relative to reference	Emissions still increasing, policies under development

Melting Glaciers and Rising Sea Levels: One of the most dramatic and direct impacts of global warming is the rapid melting of glaciers in Polar Regions and mountainous areas. Rising global temperatures are causing significant ice mass loss in both large ice sheets—such as Greenland and Antarctica—and regional glaciers in the Alps, Himalayas, and Andes. This process directly contributes to sea level rise, while also disrupting hydrological cycles, climate systems, and ecosystem balances. Over the past 30 years, the global sea level has risen by approximately 10 cm, and this increase continues to accelerate (NASA, 2022).

Global Warming and Glacier Mass Loss

- Over the past century, the rise in global average temperatures has led to large-scale melting of glacier masses.
- Between 2000 and 2019, mountain glaciers worldwide lost an average of 267 billion tons of ice per year (Hugonnet et al., 2021).
- The Greenland Ice Sheet alone lost an average of 279 billion tons of ice per year between 2011 and 2020, contributing approximately 0.77 mm per year to global sea level rise (IPCC, 2021). In 2021 alone, Greenland lost 40 billion tons of ice, equivalent to the annual water consumption of the entire United States (NASA, 2022).
- The Antarctic Ice Sheet, during the same period, experienced an average annual mass loss of 148 billion tons, contributing an additional 0.42 mm per year to sea level rise (Rignot et al., 2019).

Sea Level Rise: Numerical Facts

- Between 1901 and 2018, the global mean sea level rose by approximately 20 cm (IPCC, 2021).

However, the rate of increase has accelerated significantly in recent decades:

- 1993–2002: Average annual rise of 2.1 mm
- 2006–2018: Average annual rise of 3.7 mm (IPCC, 2021)

By the year 2100, projected sea level rise is as follows:

- Optimistic scenario (SSP1-1.9): +0.29 m
- Intermediate scenario (SSP2-4.5): +0.44 m
- Pessimistic scenario (SSP5-8.5): +0.77 m (or higher) (IPCC, 2021)

Global and Regional Impacts

- Rising sea levels pose a direct threat to hundreds of millions of people living in coastal areas. By 2100, nearly 200 million people are expected to be exposed to coastal flooding (Kulp & Strauss, 2019).
- Approximately 10% of the world's population (around 800 million people) live at elevations of 10 meters or less above sea level.
- Countries such as Bangladesh, the Netherlands, the Maldives, and Indonesia are among those at highest risk. For instance, Bangladesh faces the threat of losing 17% of its land area if sea levels rise by around one meter — a scenario that would displace millions of people (IPCC, 2022).
- Under the pessimistic scenario, Europe is expected to experience the most significant climatic shift among continents, with 65–91% of its land projected to fall within a different climate zone by the end of the century (Bayar et al., 2023). Particularly, the expansion of the temperate climate zone into areas that currently have cold climates is anticipated to become evident by the late 21st century.
- In Türkiye, projected sea level changes related to climate change indicate the emergence of high-risk coastal areas. Assessments reveal that Adana, Edirne, İzmir, and Samsun are expected to be the provinces most affected by sea level rise (Geymen & Dirican, 2016).

Irreversible Threshold: Trigger Mechanisms in Glacier Systems

- Scientists express concern about irreversible melting thresholds, particularly for the Greenland and West Antarctic ice sheets. A global temperature increase of 1.5–2°C could cause these thresholds to be exceeded (Pattyn et al., 2018).
- If these tipping points are crossed, sea level rise could continue for centuries, ultimately leading to an increase of 2 to 7 meters worldwide.

Extreme Weather Events: Frequency, Intensity, and Climate Linkages

Climate change has increased the frequency, intensity, and duration of extreme weather events globally. These include heatwaves, droughts, floods, wildfires, storms, and tropical cyclones, which not only cause severe human and economic losses, but also exert devastating effects on ecosystems and biodiversity (IPCC, 2021).

In 2023, India experienced a heatwave reaching 50°C, Europe faced combined drought and wildfires, and Canada suffered the largest wildfire season in its recorded history. The extreme weather conditions that triggered recent deadly wildfires — including those affecting Türkiye — are now estimated to be ten times more likely due to climate change (World Weather Attribution [WWA], 2025). These phenomena are direct manifestations of the ongoing climate crisis.

Heatwaves and Record Temperatures

- The summer of 2023 was the hottest ever recorded in Europe (C3S, 2023).
- Record temperatures were measured in Greece (47°C), Italy (48.2°C), and Spain (45°C) during 2023.
- In 2023, the world experienced 86 days that were more than 1.5°C warmer than the pre-industrial average (WMO, 2024).
- Over the last 20 years, the frequency of heatwaves has tripled, while their average duration has doubled (Perkins-Kirkpatrick & Lewis, 2020).
- In Türkiye, the highest temperature on record was 49.5°C, measured on 15 August 2023 in Sarıcakaya, Eskişehir (MGM, 2025).

- Under current policies, with projected warming reaching 2.6°C this century, extremely dry conditions are expected to become six times more likely compared to today (WWA, 2025).

Floods and Extreme Precipitation Events

- Due to global warming, each 1°C rise in atmospheric temperature enables the air to hold about 7% more water vapor, intensifying precipitation events (Trenberth, 2011).
- In 2022, severe flooding in Pakistan submerged 33% of the country, affected 33 million people, and caused the deaths of over 1,700 individuals (UN OCHA, 2022).
- In Germany and Belgium, the 2021 summer floods caused over 220 fatalities and resulted in €30 billion in damages (EM-DAT, 2021).
- In Türkiye, the highest daily rainfall was recorded on 17 December 2018, reaching 490.8 mm in Kemer-Ovacık, Antalya (MGM, 2025).
- On 11 August 2021, severe rainfall in the Western Black Sea region of Türkiye triggered floods and landslides, resulting in 97 deaths and 228 injuries across Kastamonu, Sinop, and Bartın.
- On 15 March 2023, extreme rainfall in southeastern Türkiye led to devastating floods, primarily affecting Adıyaman and Şanlıurfa, two of the country's driest regions. The disaster caused 21 deaths and damaged 3,154 homes, businesses, vehicles, and properties.

Droughts and the Water Crisis

- Between 2020 and 2022, East Africa experienced its most severe drought in 40 years, leaving approximately 20 million people in a state of food insecurity (World Food Programme [WFP], 2023).
- The European drought of summer 2022 was the worst in 500 years, leading to significantly reduced river flows and agricultural productivity losses of 10–20% on average (C3S, 2022).
- In Türkiye, during the autumn of 2023, water levels in Istanbul's reservoirs dropped to below 20% capacity (İSKİ, 2023).

Forest Fires

- The year 2025 has been reported as the worst wildfire year ever recorded in Europe, with over one million hectares of land burned while record-breaking temperatures were observed in Türkiye, Cyprus, Spain, France, and the United Kingdom (WWA, 2025).
- The summer of 2022 was the hottest on record in Europe. Thousands of hectares of forests were destroyed in France, Spain, Portugal, and Greece. These fires were primarily the result of extreme drought and heatwaves (Copernicus, 2023).
- In 2023, wildfires in Canada burned 18 million hectares, marking the largest fire season in the nation's history (Natural Resources Canada, 2023).
- During the 2019–2020 Australian wildfires, it is estimated that three billion animals were killed or displaced (WWF, 2020).
- In Türkiye, wildfires during the summer of 2021 burned approximately 140,000 hectares of land, recorded as Europe's largest forest loss event (TEMA, 2021).
- In 2022, 2,160 wildfires occurred in Türkiye, burning 12,799 hectares of forest. Between January 1 and August 19, 2023, 1,419 wildfires were recorded, while in the same period of 2024, the number increased to 2,529 fires. The burned forest area also rose from 8,865 hectares to 17,456 hectares over the same period.

Türkiye – Forest Fires (2005–2025)

Year (Period)	Number of Fires	Burned Area (ha)	Source
2005–2024 Average	2,732 fires/year	25,762 ha/year	Turkish Foresters Association (2025)
2021 (Record Year)	2,793	139,503 ha	Turkish Foresters Association (2025); *Tarım Dünyası* (2025)
2024	3,797	27,485 ha	Anadolu Agency (2025); *Tarım Dünyası* (2025)
2025 (Jan 1 – July 30)	4,426	49,769 ha	Kocaeli Fikir (2025)

The general trend indicates a significant increase in both the number of fires and the total burned area, especially after 2020. Regionally, the Aegean and Mediterranean regions are the most affected, with Muğla, Antalya, and İzmir being the leading provinces. The summer of 2025 (especially from early July onward) witnessed a notable escalation in fire incidents across many regions and provinces compared to previous years. This surge appears to have been further triggered by climate-related factors, as well as human-induced causes, including negligence and arson.

Tropical Storms and Hurricanes

- Although there has been no significant increase in the overall number of tropical storms worldwide, the proportion of Category 4 and 5 storms has increased by 25% over the last 40 years (Knutson et al., 2020).
- In 2022, Hurricane Ian in the United States caused economic losses of USD 112 billion, making it the most devastating event of the year (NOAA, 2023).

Economic and Human Costs

Between 2000 and 2019, climate-related disasters resulted in approximately:

- 7.3 billion People affected,
- 1.23 million Deaths, and
- USD 2.97 trillion in economic losses (UNDRR, 2020).

The economic cost of extreme weather events alone in 2023 was estimated at around USD 360 billion (Swiss Re Institute, 2024).

Agriculture and Food Crisis: A Global Threat Deepened by Climate Change

Climate change is one of the most significant environmental threats directly affecting agricultural production, food security, and food safety. Rising temperatures, shifting precipitation patterns, the increase in extreme weather events (droughts, floods, hail, storms, frost, etc.), and the spread of pests and harmful organisms reduce agricultural productivity, leading to higher food prices and food insecurity (Aydogdu & Kaya, 2020; IPCC, 2022).

Impact of Climate Change on Agricultural Yields

According to the IPCC (2022) assessment, during the 1981–2010 period:

- Wheat production was negatively affected by approximately 6%,
- Maize (corn) production by 7.4%,
- Rice production by 3.2% due to global temperature rise.
- By 2050, temperature increases and water stress may reduce wheat yields by up to 20% (FAO, 2021).
- Every 1°C rise in temperature could cause more than a 10% decline in maize and rice yields in tropical regions (Zhao et al., 2017).

Drought and Water Scarcity

- Agricultural irrigation accounts for approximately 70% of global water use by sector (Aydogdu & Yenigün, 2016; FAO, 2022).
- However, climate change is rapidly depleting groundwater in many regions and altering precipitation patterns.
- For example, in Southeastern Türkiye in 2023, rainfall was 45% below the long-term average, leading to significant yield reductions in cereal crops (TÜİK, 2023).

Impact of Extreme Weather Events on Agriculture

- During the period 2005–2015, climate-related disasters caused approximately USD 96 billion in economic losses in the agricultural sector of developing countries (FAO, 2018).
- In 2021, drought in the United States caused USD 1.1 billion in damages to California's agricultural sector alone (USDA, 2022).
- In Türkiye, frost events in the summer of 2021 caused up to 60% losses in apricot and grape production (Ministry of Agriculture and Forestry, 2021).

Food Security and Hunger Risks

- According to FAO (2023), 735 million people worldwide are undernourished.
- This represents an increase of approximately 122 million people compared to 2019, with climate change, pandemics, and conflicts being major contributing factors.
- By 2050, the global population is expected to reach 9.7 billion, and agricultural production may fail to meet demand due to climate impacts (FAO, 2021).
- The IPCC (2022) predicts that even under a 1.5°C warming scenario, an additional 100 million people could face food insecurity.

Climate Change and Shifts in Agricultural Geography

- Rising temperatures are causing some crops to shift northward or to higher altitudes, threatening ecosystem balance, local economies, and biodiversity.
- For example, in Türkiye, hazelnut cultivation is moving from the traditional Black Sea coastal region toward inland areas (TAGEM, 2023).

Drought, Water Scarcity, Agricultural Collapse, and Security Implications

The depletion of water resources can trigger conflicts, cross-border crises, and civil wars. Examples include:

- In Syria (2006–2010), severe drought forced 1.5 million farmers to abandon agriculture, leading to social unrest in cities and ultimately contributing to the civil war (Kelley et al., 2015).
- In Sahel countries (including Chad, Niger, Mali, stretching from North Senegal, Southern Mauritania, the large bend of the Niger River in Mali, Burkina Faso, Southern Niger, Northeastern Nigeria, Central-Southern Chad, to Sudan), droughts and loss of pastures have facilitated the spread of armed groups and radicalization (UNDP, 2022).

- According to the IPCC, exceeding the 1.5°C threshold is expected to increase stress on water resources by more than 50%, potentially triggering security threats (IPCC, 2022).

Migration and Security: In the Context of Climate Change

Climate change is causing people to leave their homes due to water scarcity, drought, sea level rise, declines in agricultural production, and the increasing frequency of natural disasters, leading to migration waves that pose security risks both regionally and globally. This situation threatens not only physical security but also food, water, energy, and social stability (IPCC, 2022).

Climate Migration in Numbers

- In 2022, approximately 32.6 million people worldwide were displaced due to climate-related disasters (floods, storms, droughts, fires, etc.). This number is more than double those displaced by conflict (IDMC, 2023).
- During the period 2020–2022 alone, the number of people forced to migrate due to climate events exceeded 90 million (IOM, 2023).
- According to IPCC (2022) projections, by 2100, up to 1 billion people may be forced to leave regions rendered uninhabitable by climate change.

According to the World Bank (2021):

- By 2050, 216 million people in Africa, South Asia, and Latin America will be at risk of climate-induced migration.
- In Sub-Saharan Africa alone, this figure could reach 86 million.

Sea Level Rise and Island States

Global sea level has risen by approximately 20 cm since 1901, and by 2100, it is expected to rise between 0.5 and 1 meter (IPCC, 2021).

This rise could result in:

- 20 million people in Bangladesh being displaced,
- 17 million people in Vietnam,

- Nearly entire populations of island states such as the Maldives (Kulp & Strauss, 2019).
- Countries like Tuvalu, Kiribati, and the Marshall Islands are leading mass migration waves, potentially becoming the first "climate refugees" (McNamara & Farbotko, 2017).

Climate Security: International Risks and Policy Impact

- International organizations such as NATO, the EU, and the UN now recognize climate change as a "threat multiplier."
- The Pentagon considers climate change a national security risk (U.S. Department of Defense, 2021).
- In 2021, the UN Security Council issued a comprehensive report noting that climate change increases conflict risk and hinders peace (UNSC, 2021).
- The status of climate migrants remains uncertain under international law, making the issue of "climate refugees" legally, morally, and politically complex (Biermann & Boas, 2010).

The Case of Türkiye and Regional Dynamics

- Türkiye is both a target and transit country for climate-induced migration, being exposed to the destructive effects of climate change. Its historical size, geopolitical significance, and deeply rooted social values of compassion and solidarity make it a reliable refuge during such crises.
- Climate crises in the Middle East and Africa impact Türkiye's internal and external migration dynamics.
- In Türkiye, particularly in Southeastern Anatolia and Central Anatolia, droughts and agricultural income losses are driving rural-to-urban migration, exacerbating urban infrastructure and social integration challenges.

Chapter Conclusion

The multidimensional crises discussed in this section remind us that the environment and nature are not just a habitat but a dynamic, fragile, and finite system. Climate change is not only an environmental issue but also a social, economic, and political challenge. From carbon emissions to food security, from consumption patterns to migration flows, the reactions of nature and natural resources can no longer be ignored.

It is essential to view these crises holistically rather than as isolated events. Reducing fossil fuel use, promoting sustainable agriculture, transforming consumption habits, and ensuring climate justice are not only environmental responsibilities but moral obligations.

The data presented throughout this section show that nature and natural resources act not as silent observers but as warning mechanisms. Taking these warnings seriously and acting upon them is crucial for the survival of humanity.

Nature and its resources are living, responsive systems, not passive. The storms brought by wind, drying soils, receding waters, increasing fires, and mass migrations are the rebellions of nature and natural resources. These are not acts of anger or revenge; they are humanity's final call: "Stop and listen" and "Show care and sensitivity." Only then can positive change begin.

Nature and resources remind us of their limits. Their resistance against depletion is also a warning against destruction. We must transform our consciousness, production, consumption, and relationships. This rebellion is essentially a call for justice and peace. If humanity answers this call, nature, the environment, and natural resources will again reveal their compassionate faces. Otherwise, nature's justice will be harsher, more certain, and irreversible than any human-centered construct. Nature is no longer silent; it is speaking in a language humans can understand. Listening and responding is now our responsibility.

Chapter References

AMAP. (2021). *Arctic Climate Change Update 2021: Key Trends and Impacts.* Arctic Monitoring and Assessment Programme Retrieved from. https://www.amap.no

Anadolu Ajansı. (2025, Ocak 15). *Türkiye'de 240 bin hektar ormanlık alan yandı iddiası* [Claim that 240 thousand hectares of forest area burned in Türkiye]. Retrieved from https://www.aa.com.tr/tr/teyithatti/aktuel/turkiyede-240-bin-hektar-ormanlik-alan-yandi-iddiasi/1819350

AURA. (2025). *Sera gazı nedir, sera gazları nelerdir, nasıl oluşur, sera gazı etkisi nedir, sera gazı emisyonu nedir, nasıl azaltılır?* [What is a greenhouse gas, what are greenhouse gases, how do they form, what is the greenhouse effect, what is greenhouse gas emission, how can it be reduced?]. Aura Belgelendirme. Retrieved from https://www.seragazidogrulama.com/sera-gazi-nedir-sera-gazlari-nelerdir-nasil-olusur-sera-gazi-etkisi-nedir-sera-gazi-emisyonu-nedir-nasil-azaltilir

Aydogdu, M. H., & Kaya, F. (2020). Factors affecting consumers' consumption of organic foods: A case study in GAP-Şanlıurfa in Turkey. *Journal of Agricultural Science and Technology, 22*(2), 347–359. Retrieved from http://journals.modares.ac.ir/article-23-13340-en.html

Aydogdu, M. H., & Yenigün, K. (2016). Farmers' risk perception towards climate change: A case of the GAP–Şanlıurfa region, Turkey. *Sustainability, 8*(8), 806. https://doi.org/10.3390/su8080806

Aydoğdu, M. H., Sevinç, M. R., & Cançelik, M. (2021). Determination of farmers' willingness to pay for drought adaptation policies in Şanlıurfa, Turkey. *Weather, Climate, and Society, 13*(3), 677–686. https://doi.org/10.1175/WCAS-D-20-0163.1

Ballester, J., Scortichini, M., Russo, S., Sera, F., Gasparrini, A., & Vicedo-Cabrera, A. M. (2023). The impact of the 2022 summer heatwave on excess mortality in Europe. *Nature Medicine, 29*(8), 1825–1834. https://doi.org/10.1038/s41591-023-02554-3

Bayar, A. S., Yılmaz, M. T., Yücel, İ., & Dirmeyer, P. (2023). CMIP6 Earth system models project greater acceleration of climate zone change due to stronger warming rates. *Earth's Future, 11*, e2022EF002972. https://doi.org/10.1029/2022EF002972

Beck, H. E., Lange, S., & Ciais, P. (2024). Anthropogenic contributions to heat-related mortality during the 2022 European heatwave. *Environmental Research Letters, 19*(4), 044002. https://doi.org/10.1088/1748-9326/ad1a11

Biermann, F., & Boas, I. (2010). Preparing for a warmer world: Towards a global governance system to protect climate refugees. *Global Environmental Politics, 10*(1), 60–88. https://doi.org/10.1162/glep.2010.10.1.60

Bonan, G. B. (2008). Forests and climate change: Forcings, feedbacks, and the climate benefits of forests. *Science, 320*(5882), 1444–1449. https://doi.org/10.1126/science.1155121

C3S (Copernicus Climate Change Service). (2022). *European State of the Climate 2022.* Retrieved from https://climate.copernicus.eu/esotc/2023

C3S (Copernicus Climate Change Service). (2023). *Summer 2023: Warmest on record.* Retrieved from https://climate.copernicus.eu

Chakravarty, S., Chikkatur, A., de Coninck, H., Pacala, S., Socolow, R., & Tavoni, M. (2009). Sharing global CO_2 emission reductions among one billion high emitters. *Proceedings of the National Academy of Sciences, 106*(29), 11884–11888. https://doi.org/10.1073/pnas.0905232106

Climate Action Tracker. (2023). *Warming Projections Global Update.* Retrieved from https://climateactiontracker.org

Creutzig, F., Jochem, P., Edelenbosch, O. Y., Mattauch, L., van Vuuren, D. P., McCollum, D., & Minx, J. (2015). Transport: A roadblock to climate change mitigation? *Science, 350*(6263), 911–912. https://doi.org/10.1126/science.aac8033

Dinçer, S., & Akın, F. (2025). Tarımsal toplam faktör verimliliği ve ekonomik büyümenin çevre kalitesi üzerindeki etkisi: Tarımda başarılı olan ülkeler örneği [The impact of agricultural total factor productivity and economic growth on environmental quality: A case study of successful countries in agriculture]. *Kahramanmaraş Sütçü İmam Üniversitesi Tarım ve Doğa Dergisi, 28*(6), 1531–1544. https://doi.org/10.18016/ksutarimdoga.vi.1661043

Ellen MacArthur Foundation. (2017). *A new textiles economy: Redesigning fashion's future.* Retrieved from https://ellenmacarthurfoundation.org/a-new-textiles-economy

EM-DAT. (2021). *The International Disaster Database.* Centre for Research on the Epidemiology of Disasters (CRED). Retrieved from https://www.emdat.be

European Centre for Medium-Range Weather Forecasts/Climate Change Service. (2025). *European State of the Climate 2024.* Copernicus Climate Change Service. Retrieved from https://climate.copernicus.eu

European Environment Agency (EEA). (2021). *Transport and environment report 2021: Decarbonising Europe's transport.* Retrieved from

https://www.eea.europa.eu

FAO. (2018). *The impact of disasters and crises on agriculture and food security.* Food and Agriculture Organization of the United Nations. Retrieved from https://www.fao.org

FAO. (2019). *The state of food and agriculture: Moving forward on food loss and waste reduction.* Food and Agriculture Organization of the United Nations. Retrieved from https://www.fao.org

FAO. (2020). *Global forest resources assessment 2020.* Food and Agriculture Organization of the United Nations. Retrieved from https://www.fao.org/forest-resources-assessment

FAO. (2020). *Greenhouse gas emissions from agriculture, forestry and other land use.* Food and Agriculture Organization of the United Nations. Retrieved from https://www.fao.org

FAO. (2021). *Climate-smart agriculture and the future of food.* Food and Agriculture Organization of the United Nations. Retrieved from https://www.fao.org

FAO. (2021). *The impact of disasters and crises on agriculture and food security: 2021.* Food and Agriculture Organization of the United Nations. Retrieved from https://openknowledge.fao.org/items/e5a5771a-6127-4e4a-ae1c-9f26fe8aa770

FAO. (2022). *FAO statistical yearbook 2022.* Food and Agriculture Organization of the United Nations. Retrieved from https://www.fao.org

FAO. (2023). *The state of food security and nutrition in the world 2023.* Food and Agriculture Organization of the United Nations. Retrieved from https://www.fao.org

Gatti, L. V., Basso, L. S., Miller, J. B., et al. (2021). Amazonia as a carbon source linked to deforestation and climate change. *Nature, 595*(7867), 388–393. https://doi.org/10.1038/s41586-021-03629-6

Gerber, P. J., Steinfeld, H., Henderson, B., et al. (2013). *Tackling climate change through livestock: A global assessment of emissions and mitigation opportunities.* Food and Agriculture Organization of the United Nations. Retrieved from https://www.fao.org/3/i3437e/i3437e.pdf

Geymen, A., & Dirican, A. Y. (2016). İklim değişikliğine bağlı deniz seviyesi değişiminin coğrafi bilgi sistemleri kullanılarak analiz edilmesi [Analysis of sea level changes related to climate change using geographic information systems]. *Harita Teknolojileri Elektronik Dergisi, 8*(1), 65–74. https://doi.org/10.15659/hartek.16.04.308

Global Forest Watch. (2023). *Forest loss data.* World Resources Institute. Retrieved from https://www.globalforestwatch.org

Gössling, S., Humpe, A., & Bausch, T. (2020). Does 'flight shame' affect social norms? Changing perspectives on the desirability of air travel in Germany. *Journal of Cleaner Production, 266,* 121970. https://doi.org/10.1016/j.jclepro.2020.121970

Griscom, B. W., et al. (2017). Natural climate solutions. *Proceedings of the National Academy of Sciences, 114*(44), 11645–11650. https://doi.org/10.1073/pnas.1710465114

Gülersoy, A. E., Kaya, G., & Şeker, A. (2024). Çevre ve toplum [Environment and society]. Retrieved from https://www.researchgate.net/publication/380792354_CEVRE_VE_TOPLUM

Gültekin, S. (2024). Çevre sosyolojisine kültürel bir bakış [A cultural perspective on environmental sociology]. https://doi.org/10.62147/veche.1495672

Hoegh-Guldberg, O., Jacob, D., Taylor, M., et al. (2018). *Impacts of 1.5°C global warming on natural and human systems.* In: *IPCC Special Report on Global Warming of 1.5°C.* Retrieved from https://www.ipcc.ch/sr15/

Hugonnet, R., McNabb, R., Berthier, E., et al. (2021). *Accelerated global glacier mass loss in the early twenty-first century. Nature*, 592, 726–731. https://doi.org/10.1038/s41586-021-03436-z

IDMC. (2023). *Global Report on Internal Displacement 2023.* Internal Displacement Monitoring Centre. Retrieved from https://www.internal-displacement.org

IEA. (2021). *Global Energy Review: CO_2 Emissions in 2021.* International Energy Agency. Retrieved from https://www.iea.org/reports

IEA. (2022). *CO_2 Emissions from Fuel Combustion 2022.* International Energy Agency. Retrieved from https://www.iea.org/reports

IOM. (2023). *World Migration Report 2023.* International Organization for Migration. Retrieved from https://www.iom.int

IPCC. (2021). *Climate Change 2021: The Physical Science Basis.* Contribution of Working Group I to the Sixth Assessment Report of the Intergovernmental Panel on Climate Change. Retrieved from https://www.ipcc.ch/report/ar6/wg1/

IPCC. (2022). *Climate Change 2022: Impacts, Adaptation and Vulnerability.* Intergovernmental Panel on Climate Change. Retrieved from https://www.ipcc.ch/report/ar6/wg2/

Ivanova, D., Stadler, K., Steen-Olsen, K., Wood, R., Vita, G., Tukker, A., & Hertwich, E. G. (2016). *Environmental impact assessment of household consumption. Journal of Industrial Ecology*, 20(3), 526–536. https://doi.org/10.1111/jiec.12371

İSKİ. (2023). *Baraj Doluluk Oranları (Dam Occupancy Rates).* İstanbul Su ve Kanalizasyon İdaresi (Istanbul Water and Sewerage Administration). Retrieved from https://www.iski.istanbul

Jackson, T. (2009). *Prosperity Without Growth: Economics for a Finite Planet.* Earthscan, London, UK. Retrieved from https://www.environmentandsociety.org/mml/prosperity-without-growth-economics-finite-planet

Kelley, C. P., Mohtadi, S., Cane, M. A., Seager, R., & Kushnir, Y. (2015). *Climate change in the Fertile Crescent and implications of the recent Syrian drought. PNAS*, 112(11), 3241–3246. https://doi.org/10.1073/pnas.1421533112

Knutson, T. R., Camargo, S. J., Chan, J. C. L., et al. (2020). *Tropical cyclones and climate change assessment. Bulletin of the American Meteorological Society*, 101(3), E303–E322. https://doi.org/10.1175/BAMS-D-18-0189.1

Kocaeli Fikir. (2025). *2025 yazı Türkiye'de kaç orman yangını çıktı? (How many forest fires occurred in Turkey in the summer of 2025?)* Retrieved from https://www.kocaelifikir.com/foto-galeri/166538-2025-yazi-turkiyede-kac-orman-yangini-cikti

Kulp, S. A., & Strauss, B. H. (2019). *New elevation data triple estimates of global vulnerability to sea-level rise and coastal flooding. Nature Communications*, 10, 4844. https://doi.org/10.1038/s41467-019-12808-z

Kurnaz, M. L. (2023). *İklim Değişikliği ve Uyum Süreçlerinde Türkiye (Climate Change and Adaptation Processes in Turkey). Dirençlilik Dergisi (Resilience Journal)*, 7(1), 199–208. https://doi.org/10.32569/resilience.1312684

Malhi, Y., Roberts, J. T., Betts, R. A., et al. (2008). *Climate change, deforestation, and the fate of the Amazon. Science*, 319(5860), 169–172. https://doi.org/10.1126/science.1146961

McNamara, K. E., & Farbotko, C. (2017). *Resisting relocation: Place attachment in climate-affected communities in Fiji. Forced Migration Review*, (55), 64–66.

MGM. (2022). *Türkiye İklim Raporu 1970–2020 (Turkey Climate Report 1970–2020).* Meteoroloji Genel Müdürlüğü (Turkish State Meteorological Service). Retrieved from https://mgm.gov.tr

MGM. (2024). *2023 Yılı Türkiye İklim Değerlendirme Raporu (2023 Turkey Climate Assessment Report).* Meteoroloji Genel Müdürlüğü (Turkish State Meteorological Service). Retrieved from https://mgm.gov.tr

MGM. (2025). *Türkiye'de ve Dünyada Kaydedilen En Düşük ve En Yüksek Değerler (The Lowest and Highest Recorded Values in Turkey and Worldwide).* Meteoroloji Genel Müdürlüğü (Turkish State Meteorological Service). Retrieved from https://www.mgm.gov.tr/genel/sss.aspx?s=sicaklikenleri2

NASA. (2022). *Vital Signs of the Planet: Ice Sheets.* Retrieved from https://climate.nasa.gov

NASA. (2024). *Global Climate Change: Vital Signs of the Planet.* Retrieved from https://climate.nasa.gov

Natural Resources Canada. (2023). *National Wildland Fire Situation Report.* Retrieved from https://cwfis.cfs.nrcan.gc.ca

NOAA. (2023). *Billion-Dollar Weather and Climate Disasters.* Retrieved from https://www.ncei.noaa.gov

OGM. (2022). *Türkiye Orman Varlığı 2021 (Turkey Forest Inventory 2021).* Tarım ve Orman Bakanlığı (Ministry of Agriculture and Forestry). Retrieved from https://www.ogm.gov.tr/tr

Okumuş, N. (2025). *AB üye ülkelerinin Çevre Ekonomisi Stratejilerinin Karşılaştırılması için Yeni Bir Yaklaşım (A New Approach for Comparing Environmental Economy Strategies of EU Member States). Kahramanmaraş Sütçü İmam Üniversitesi Sosyal Bilimler Dergisi*, 22(2), 1023–1036. https://doi.org/10.33437/ksusbd.1659012

Ormancılar Derneği. (2025). *Orman yangınları istatistikleri (Forest Fire Statistics).* Retrieved from https://www.ormancilardernegi.org/Yangin

Özcan, İ. (2024). *Proaktif Çevre Yönetimi Kapsamında Karbon Muhasebesi (Carbon Accounting within Proactive Environmental Management).* Retrieved from https://www.researchgate.net/publication/386445853_PROAKTIF_CEVRE_YONETIMI_KAPSAMINDA_KARBON_MUHASEBESI

Öztürk, F. N., & Özgür, Ö. (2025). *Türkiye'deki Çevre Krizlerine Sosyal Hizmet Perspektifinden Yeni Bir Bakış: Yeşil Protokoller (A New Social Work Perspective on Environmental Crises in Turkey: Green Protocols). Samsun Sağlık Bilimleri Dergisi*, 10(2), 157–172. https://doi.org/10.47115/jshs.1667982

Pan, Y., Birdsey, R. A., Fang, J., et al. (2011). *A large and persistent carbon sink in the world's forests. Science*, 333(6045), 988–993. https://doi.org/10.1126/science.1201609

Parlakçı Doğan, H., Aydogdu, M. H., Sevinç, M. R., & Cançelik, M. (2020). *Farmers' Willingness to Pay for Services to Ensure Sustainable Agricultural Income in the GAP-Harran Plain, Sanlıurfa, Turkey. Agriculture*, 10, 152. https://doi.org/10.3390/agriculture10050152

Pattyn, F., Ritz, C., Hanna, E., et al. (2018). *The Greenland and Antarctic ice sheets under 1.5 °C global warming. Nature Climate Change*, 8(12), 1053–1061. https://doi.org/10.1038/s41558-018-0305-8

Perkins-Kirkpatrick, S. E., & Lewis, S. C. (2020). *Increasing trends in regional heatwaves. Nature Communications*, 11, 3357. https://doi.org/10.1038/s41467-020-16970-7

Reisinger, A., Clark, H., et al. (2021). *How much can dietary change and livestock management reduce methane emissions? Global Change Biology*, 27(9), 2058–2075. https://doi.org/10.1111/gcb.15559

Rignot, E., Mouginot, J., Scheuchl, B., et al. (2019). *Four decades of Antarctic Ice Sheet mass balance from 1979–2017. Proceedings of the National Academy of Sciences*, 116(4), 1095–1103. https://doi.org/10.1073/pnas.1812883116

Rockström, J., & Gaffney, O. (2021). *Breaking Boundaries: The Science of Our Planet.* Penguin. Westminster, London, UK.

Saunois, M., Stavert, A. R., Poulter, B., et al. (2020). *The Global Methane Budget 2000–2017. Earth System Science Data*, 12(3), 1561–1623. https://doi.org/10.5194/essd-12-1561-2020

Sevinç, G., Aydogdu, M. H., Cançelik, M., & Sevinç, M. R. (2019). *Farmers' Attitudes toward Public Support Policy for Sustainable Agriculture in GAP-Şanlıurfa, Turkey. Sustainability*, 11(23), 6617. https://doi.org/10.3390/su11236617

Swiss Re Institute. (2024). *Natural catastrophes in 2023: A record year of losses.* Retrieved from https://www.swissre.com

TAGEM. (2023). *Türkiye Tarımsal Araştırmalar ve Politikalar Genel Müdürlüğü Raporu (General Directorate of Agricultural Research and Policies Report).* Retrieved from https://www.tarimorman.gov.tr/TAGEM

Tarım Dünyası. (2025). *Orman yangınları ve felaketler geçici değil (Forest Fires and Disasters Are Not Temporary).* Retrieved from https://www.tarimdunyasi.net/2025/07/29/orman-yanginlari-ve-felaketler-gecici-degil

Tarım ve Orman Bakanlığı. (2021). *Tarımda İklim Riskleri Raporu (Climate Risk Report in Agriculture).* Retrieved from https://www.tarimorman.gov.tr

TEMA. (2021). *Yangınların Ardından Türkiye Orman Gerçeği Raporu (The Reality of Turkey's Forests after Fires).* TEMA Vakfı (TEMA Foundation).

Tolunay, D. (2025). *Ekosistem Temelli Afet Risk Azaltma, ÇEM Hidrometerolojik Afetler Eğitimi (Ecosystem-Based Disaster Risk Reduction, Hydrometeorological Disaster Training).* Kastamonu.

Trenberth, K. E. (2011). *Changes in precipitation with climate change. Climate Research*, 47, 123–138. https://doi.org/10.3354/cr00953

Trenberth, K. E., Fasullo, J. T., & Shepherd, T. G. (2014). *Attribution of climate extreme events. Nature Climate Change*, 5(8), 725–730. https://doi.org/10.1038/nclimate2657

Tuna, A., & Güzel, H. Y. (2024). *Çevre Yanlısı Davranışlar: Hangi Davranışlar Hangi Amaçlarla Gerçekleştiriliyor? (Environmentally Friendly Behaviors: Which Behaviors Are Carried Out for What Purposes?)* Retrieved from https://www.researchgate.net/publication/394845612_Cevre_Yanlisi_Davranislar_Hangi_Davranislar_Hangi_Amaclarla_Gerceklestiriliyor

TÜİK. (2022). *Sera Gazı Emisyon İstatistikleri, 2021 (Greenhouse Gas Emission Statistics, 2021).* Türkiye İstatistik Kurumu (Turkish Statistical Institute). Retrieved from https://data.tuik.gov.tr

TÜİK. (2023). *Bitkisel Üretim İstatistikleri 2023 (Crop Production Statistics 2023).* Türkiye İstatistik Kurumu. Retrieved from https://data.tuik.gov.tr

TÜİK. (2023). *Motorlu Kara Taşıtları, 2023 (Motor Vehicles, 2023).* Türkiye İstatistik Kurumu. Retrieved from https://data.tuik.gov.tr

TÜİK. (2025). *İstatistiklerle Türkiye, 2024 (Turkey in Statistics, 2024).* Türkiye İstatistik Kurumu. ISBN 978-625-5591-10-4. Retrieved from https://biruni.tuik.gov.tr/yayin/views/visitorPages/yayinGoruntuleme.zul?yayin_no=683

U.S. Department of Defense. (2021). *Climate Risk Analysis.* Retrieved from https://www.defense.gov

UN OCHA. (2022). *Pakistan Floods Situation Report.* Retrieved from https://reliefweb.int

UNDP. (2022). *Human Development Report 2022.* Retrieved from https://www.undp.org

UNDRR. (2020). *The Human Cost of Disasters: An Overview of the Last 20 Years (2000–2019).* Retrieved from https://www.undrr.org

UNEP & CCAC. (2021). *Global Methane Assessment: Benefits and Costs of Mitigating Methane Emissions.* United Nations Environment Programme and Climate and Clean Air Coalition. Retrieved from https://www.unep.org/resources/report/global-methane-assessment

UNFCCC. (2015). *Paris Agreement.* United Nations Framework Convention on Climate Change. Retrieved from https://unfccc.int

UNSC. (2021). *Climate and Security: Report of the Secretary-General to the Security Council.* Retrieved from https://www.un.org/securitycouncil

USDA. (2022). *California Drought Impact Report.* Retrieved from https://www.usda.gov

Walker, W. S., Gorelik, S. R., Cook-Patton, S. C., et al. (2020). *The role of forest conservation in mitigating climate change. Nature Climate Change*, 10(12), 1039–1046. https://doi.org/10.1038/s41558-020-00926-5

WFP. (2023). *Hunger Crisis in the Horn of Africa.* Retrieved from https://www.wfp.org

WMO. (2024). *State of the Global Climate 2023.* World Meteorological Organization. Retrieved from https://wmo.int

World Bank. (2021). *Groundswell Part 2: Acting on Internal Climate Migration.* The World Bank. Retrieved from https://openknowledge.worldbank.org/entities/publication/2c9150df-52c3-58ed-9075-d78ea56c3267

World Economic Forum. (2020). *The Future of the Last-Mile Ecosystem.* Retrieved from https://www.weforum.org/reports/the-future-of-the-last-mile-ecosystem

World Health Organization Regional Office for Europe. (2025). *Projected impacts of climate change on heat-related mortality in Europe.* WHO/Europe. Retrieved from https://www.who.int/europe

World Health Organization. (2024). *Climate change and health: Heat and health.* WHO. Retrieved from https://www.who.int/publications

WWA. (2025). *Report, 2025-08-28, Centre for Environmental Policy.* https://doi.org/10.25560/123302

WWF. (2020). *Australia's 2019–2020 Bushfires: The Wildlife Toll.* https://www.wwf.org.au

Yan, X., Akiyama, H., Yagi, K., & Akimoto, H. (2009). *Global estimations of the inventory and mitigation potential of methane emissions from rice cultivation conducted using the 2006 IPCC Guidelines. Global Biogeochemical Cycles*, 23(2). https://doi.org/10.1029/2008GB003299

Yılmaz, M. T. (2025). *İklim Değişikliğinin Afetler Üzerine Etkisi (The Impact of Climate Change on Disasters).* ÇEM Hidrometerolojik Afetler Eğitimi, Kastamonu.

Zhao, C., Liu, B., Piao, S., et al. (2017). *Temperature increase reduces global yields of major crops in four independent estimates. PNAS*, 114(35), 9326–9331. https://doi.org/10.1073/pnas.1701762114

4th CHAPTER

WATER RESOURCES, NATURAL SYSTEMS, AND ETHICAL RESPONSIBILITY: A HOLISTIC ASSESSMENT FROM A GLOBAL AND TURKISH PERSPECTIVE

"We see no essential difference between protecting our water and protecting our homeland."

— H.E. Recep Tayyip Erdoğan, President of the Republic of Türkiye

Chapter Summary

This chapter approaches water not merely as a biological necessity, but as a cultural, ethical, economic, and ecological entity. Throughout civilizations, the moral and religious sanctity attributed to water gradually gave way to a human-centered and instrumental perception, deepening global water crises. Religious traditions such as Islam and other Abrahamic faiths, as well as Buddhism, Taoism, and philosophical perspectives such as deep ecology and ecocentric ethics, emphasize that respect for water should extend beyond human benefit.

The chapter scientifically analyzes the pressures of climate change, drought, population growth, and urbanization on water resources, presenting detailed classifications of drought types and monitoring tools. At both the global and Turkish scales, it compares sectoral water use, water scarcity, efficiency, and irrigation system performance. Under the theme of "ethics of scarcity," it discusses the moral responsibility, moderation, and justice principles guiding the fair distribution of water.

***Keywords*:** Drought and Climate Change, Ethics of Scarcity and Moral Responsibility, Sectoral Water Use and Efficiency, Irrigation Systems and Water Management, Water Ethics

Water: The Heart of Natural Existence and a Moral Responsibility

Water is an indispensable element of life for all living beings on Earth (Aydogdu et al., 2015). Human intervention in the environment and nature begins with the way water is used. The way we manage natural resources reflects our relationship with nature and the environment. In this sense, as Berry (1999) states, "How we treat water is how we treat the environment and nature."

Water is not only essential for biological life but is also a divine natural entity, forming the foundation of ecosystems, social development, and human civilizations (Aydogdu et al., 2016). Throughout history, civilizations have flourished around water; river valleys have been the cradles of prosperity and agricultural societies. Rivers such as the Nile, Euphrates, Tigris, Ganges, and Yellow River have carried not only political, military, and economic significance but also religious, cultural, social, and symbolic sacred meanings (Strang, 2004).

Over time, however, this sense of sanctity has diminished. The perception of water has become increasingly anthropocentric, and it has been reduced to a mere "resource." Consequently, water has been visualized as an infinite commodity to be exploited for economic gain (Shiva, 2002). Today, humanity is facing the cost of this misconception.

Viewing water purely as an economic or technical resource neglects its divine, sacred, and natural dimensions. Across creation narratives, water has always represented the origin of life. With its healing power, its ability to restore fertility, mediate rebirth, and serve as a means of purification, water has been assigned sacred and central roles in numerous cultural and religious texts. As an indispensable component of biological life, water is vital not only for all living organisms but also for maintaining the geophysical balance of the planet.

Water in Religious and Philosophical Thought

In Islamic thought, water is regarded as the essence of life and a symbol of divine order. In the sacred texts, water is invoked not only in its physical sense but also as a metaphysical element that sustains both existence and faith. It represents the beginning of creation, the continuity of life, and the manifestation of divine mercy upon the Earth.

The Qur'an presents water as more than a biological necessity—it is a divine sign (ayah) that reveals the interconnectedness of all forms of life. Every drop that descends from the heavens is seen as part of humanity's sustenance and survival, reminding believers that water does not belong to human possession but exists under the will and governance of the Divine. Hence, the preservation of water is understood not merely as an ecological responsibility but as a spiritual duty entrusted to humankind.

Within this framework, human beings are permitted to benefit from water, yet they are not granted the right to waste, pollute, or exploit it. Water is a trust (amanah) bestowed by God, and the relationship humanity establishes with it must be governed by gratitude, moderation, and justice. Thus, in Islam, water stands both as the source of life and as a moral test of human responsibility—a sacred reminder of the balance between divine generosity and human stewardship (Diyanet İşleri Başkanlığı, 2021).

Similarly, Buddhist, Shamanic, and Taoist traditions perceive water as the spirit of life, a symbol of balance, and a path to inner purification (Chapple, 1998; Nasr, 1996). Modern environmental ethics aligns with this view.

The philosophy of deep ecology—which rejects human supremacy over nature and upholds nature's intrinsic value (Naess, 1989)—and ecocentric ethics, which attributes inherent worth to all entities, living or non-living (Attfield, 1991), both argue that ethical responsibility toward water arises not merely from human benefit, but from respect for nature itself.

Ethical Dimensions of Water Use

With climate change, population growth, expanding irrigation, development pressures, urbanization, and industrialization, water resources are increasingly strained in both quantity and quality, turning access to water into a matter of justice, morality, and politics (Gleick, 2014; Sachs, 2015; Aydogdu et al., 2018; UNESCO-WWAP, 2023).

As of 2022, 2 billion people lack access to safe drinking water, and 3.6 billion experience significant seasonal water scarcity (UNICEF & WHO, 2022). These conditions have transformed water access from a technical issue into a matter of justice and human rights (Gleick, 2014).

According to Jeffrey Sachs (2015), water is the most critical development issue of the 21st century, and water policies must consider not only economic efficiency, but also ecological balance and social inclusiveness.

Ethical Approaches to the Use of Natural Resources

Approach Type	Description	Relationship with Water and Nature
Anthropocentrism	Human-centered	Natural resources exist solely for human benefit.
Eco-centrism	Nature-centered	Water and soil have intrinsic value; balance must be preserved.
Virtue Ethics	Moderation and gratitude	Highlights consumption ethics, humility, and thankfulness.
Theo-ecology	GOD-centered ethics	Water is a divine trust; using it mindfully is an act of worship (Denny, 1998).

Water is no longer merely a resource; it has become the heart of nature, the language of life, and the test of humanity.

Global Analysis of Precipitation, Drought, and Climate Change

According to the Intergovernmental Panel on Climate Change (IPCC) Report: "Globally, imbalances in precipitation regimes have increased over the past 50 years; while some regions have experienced more intense rainfall and floods, others have faced more frequent droughts."

Global warming affects basin hydrology in a multidimensional manner. Rising temperatures increase evapotranspiration and consequently water demand, while spatial and temporal shifts in precipitation, the intensification of extreme events, and reductions in snow and ice storage further complicate water management (Ateşoğlu et al., 2025).

Trends and Changes in Precipitation

- Climate change influences precipitation not on a global uniform scale, but regionally.

- Precipitation tends to increase in tropical regions (e.g., Southeast Asia, the Amazon Basin), while it tends to decrease in subtropical and mid-latitude regions (e.g., the Mediterranean, South Africa, and inland Australia).
- Rainfall is becoming more irregular and intense, leading to increased risks of severe storms and floods.

Drought Trends

- Drought manifests in meteorological (precipitation deficiency), agricultural (low soil moisture), hydrological (surface and groundwater shortage), and socio-economic forms.
- Drought frequency and severity have increased, particularly in Africa's Sahel region, the Middle East, Southern Europe, the western United States, and Central Asia.
- During the 2020s, drought-induced migration and food crises have caused serious challenges in certain regions.

Relationship with Climate Change

- Rising temperatures → increased evaporation → loss of soil and water moisture → elevated drought risk.
- At the same time, reduced snowfall and earlier snowmelt are altering river regimes.
- Each 1°C increase in atmospheric temperature raises the likelihood of extreme precipitation and flooding events.

Analysis of Precipitation, Drought, and Climate Change in Türkiye

According to the Turkish State Meteorological Service (MGM): In 2023, total precipitation across Türkiye was 11% below the long-term average, while in Southeastern Anatolia this deficit reached nearly 30%.

Changes in Precipitation Regimes

- Türkiye lies within the Mediterranean climate zone, characterized by dry summers and wet winters.
- In recent years, total precipitation has been declining, and its seasonal distribution has become increasingly irregular:
- Winter precipitation is decreasing, while sudden downpours in spring and autumn are increasing.
- Summer drought periods are extending.
- Especially in Southeastern Anatolia and Central Anatolia, decreasing rainfall levels have reached dimensions that threaten reservoir storage capacities and agricultural production.

Between the reference period 1980–2014 and future projection periods, an average annual precipitation decrease of 3% is anticipated, although certain regions are expected to experience localized increases. In the Eastern Black Sea Region, both short- and long-term projections suggest a potential decrease in precipitation (Yılmaz, 2025).

Drought Trends in Türkiye

- Meteorological drought (precipitation deficiency) has become more frequent across the country.
- Over the past 20 years, while severe droughts previously occurred approximately once every seven years, they are now experienced every three to four years.
- Agricultural drought has deepened water scarcity, especially in agriculturally productive provinces such as Şanlıurfa, Konya, Karaman, Aksaray, and Iğdır.
- Hydrological drought is also emerging, with declining groundwater levels and reduced storage in lakes and reservoirs.

Climate Change–Related Impacts

- Temperature increase: The average temperature in Türkiye has risen by approximately 1.5°C over the past 50 years.

- Snowfall is decreasing, especially in Central and Eastern Anatolia, leading to altered hydrological regimes.
- As snow reserves diminish, spring river discharges decline, making it increasingly difficult to meet irrigation demands.
- Forest fire risk is rising due to higher temperatures and reduced humidity, particularly during summer months.

Assessment and Warnings

- Türkiye is among the most climate-vulnerable countries in the Mediterranean basin.
- Changes in precipitation regimes and droughts must be evaluated together; it is not only the quantity of water that matters, but also when and how precipitation occurs.
- There is a direct relationship between water security, food security, and energy security.
- If current trends continue, by 2040, some regions of Türkiye may transition from semi-arid to fully arid climate regimes.

Drought and Its Classifications

Drought does not merely mean "the absence of rainfall." It manifests in different forms, each leaving distinct impacts. Drought is a slow-onset natural disaster with significant ecological, agricultural, social, and economic consequences (Aydoğdu et al., 2021). The early detection and monitoring of drought have become even more crucial today, as the impacts of climate change deepen (Aydoğdu, 2024). In general, drought is classified under four main categories:

Meteorological Drought

It refers to the condition in which precipitation in a given region falls below the long-term average for normal climatic conditions.

Indicators:

- Persistent lack of rainfall, usually lasting several months.
- Begins when precipitation falls 30% below climate norms.

Impacts:

- Triggers the onset of agricultural and hydrological drought.
- Usually the first type of drought to be detected.

Agricultural Drought

Occurs when soil moisture becomes insufficient to support normal plant growth. It is influenced by factors such as precipitation deficit, high temperatures, evaporation, and wind.

Indicators:

- Decline in soil moisture levels.
- Yield losses due to plant stress.
- Significant damage occurs during drought-sensitive growth stages such as germination, flowering, and grain filling.

Impacts:

- Decreased crop yields, rural migration, increased food prices, and reductions in social welfare.

Hydrological Drought

Occurs when the water levels in rivers, lakes, and reservoirs fall below normal values. It generally emerges after meteorological drought due to a delayed effect.

Indicators:

- Reduced storage levels in lakes and reservoirs.
- Declining groundwater levels.
- Noticeable decreases in river discharge.

Impacts:

- Imbalance in sectoral water allocation, drinking water shortages, reduced energy production (especially hydropower), and irrigation water insufficiency.

Socio-Economic Drought

Refers to the impact of drought on economic, social, and political systems. It can also be viewed as the cumulative result of all other types of drought on human societies.

Indicators:

- Loss of income in agriculture and livestock sectors.
- Increases in food and water prices.
- Migration, social unrest, and rising public management costs.

Impacts:

- Rising unemployment and migration from rural areas.
- Water restrictions by municipalities, changes in government agricultural support policies, and increased agricultural debts.

Technical Classifications Based on Drought Severity

Drought Level	Description
Mild Drought	Precipitation deficit of 10–20% below normal.
Moderate Drought	Precipitation deficit of 20–40%, primarily affecting agriculture.
Severe Drought	Precipitation more than 40% below normal, with hydrological impacts observed.
Extreme Drought	Long-lasting, causing significant socio-economic impacts.

Drought Monitoring Tools

1. *Standardized Precipitation Index (SPI)*

The Standardized Precipitation Index (SPI) is a statistical indicator used to determine how much precipitation over a given period (e.g., 1, 3, 6, or 12 months) deviates from the long-term average (McKee et al., 1993).

Features:

- Based solely on precipitation data.
- Used primarily to monitor meteorological droughts.
- Applicable across various time scales — short-term for agriculture, long-term for hydrological drought.

SPI values:

- Between –1.0 and –1.5: mild drought
- Between –1.5 and –2.0: moderate drought
- Below –2.0: severe drought

The simplicity of SPI and its reliance on precipitation data make it advantageous in data-limited regions. However, it does not account for factors such as temperature or evaporation, which can limit its comprehensiveness.

2. *Standardized Precipitation–Evapotranspiration Index (SPEI)*

The SPEI functions similarly to SPI but incorporates potential evapotranspiration (PET) data, allowing it to consider temperature fluctuations and climate variability (Vicente-Serrano et al., 2010).

Features:

- Accounts for climate change and rising temperatures.
- Based on the water balance approach (precipitation minus PET).
- Suitable for applications in agriculture, hydrology, and ecosystem studies.
- Enables integrated analysis of drought and warming effects.

SPEI is a more comprehensive tool for understanding drought in regions affected by global warming, as it includes temperature-based variables that SPI excludes.

3. *Palmer Drought Severity Index (PDSI)*

The PDSI uses temperature and precipitation data to assess soil moisture and determine the intensity and duration of drought (Palmer, 1965).

Features:

- Considers temperature, precipitation, and soil type.
- –4 and below: severe drought; +4 and above: extremely wet conditions.
- Useful for agricultural and water resource management.
- Widely used in the United States but requires regional calibration for accurate application.

The PDSI is significant for linking drought to soil moisture conditions, but its performance can vary across regions if local climatic and soil parameters are not properly adapted.

4. *Drought Monitor Systems*

Developed jointly by the U.S. Department of Agriculture (USDA), NOAA, and the National Drought Mitigation Center (NDMC), this platform visualizes drought data through maps and time series, integrating meteorological, hydrological, and agricultural indicators.

Türkiye's Agricultural Drought Information System (TARBİS)

Developed by the Ministry of Agriculture and Forestry of Türkiye, TARBİS was designed to provide early monitoring, warning, and analysis of agricultural drought across the country.

Features:

- Aims for early detection of drought risk in agricultural areas.
- Integrates multiple data layers such as precipitation, temperature, humidity, and soil moisture.
- Supports TARSİM (Agricultural Insurance) policies.
- Provides province-specific risk maps and early warning mechanisms.

TARBİS plays a crucial role in protecting agricultural production and enabling timely preventive measures. Through digitalization, early warnings can now reach farmers more efficiently. Drought monitoring tools address the various dimensions of drought and serve as valuable guides for policymakers, decision-makers, farmers, and resource managers. While indices such as SPI are suitable for quick and simple applications, more advanced indices like SPEI and PDSI are preferred for analyzing complex climatic and soil conditions. In Türkiye, local systems such as TARBİS, when integrated with global models, contribute to building an effective risk management infrastructure for drought resilience.

The Global Status of Water Resources

Water Availability and Distribution

The total volume of water on Earth is approximately 1.4 billion km³, of which 97.5% is saline (in oceans and seas), and only 2.5% is freshwater. However, the amount of accessible freshwater is extremely limited, since most of it is locked in glaciers or stored underground. Specifically, 68.7% of freshwater is found in glaciers, 30.1% in groundwater, and only 1.2% in surface waters such as lakes and rivers. In other words, the freshwater directly available for human use represents merely 0.3% of the planet's total water resources (Shiklomanov, 1993). Some key figures include:

- Global annual renewable freshwater potential: 43,000 km^3 (FAO, 2021)
- Per capita water availability (2023): Global average ~5,900 m^3/year (WWAP, 2023).

Global Issues

- *Water scarcity:* Has reached critical levels, particularly in Africa, the Middle East, and South Asia.
- *Pollution:* Industrial discharges, agricultural chemicals, and domestic wastes contaminate both surface and groundwater.
- *Climate change:* Alters precipitation patterns, and increases droughts, floods, and wildfires.
- *Population growth and urbanization:* Intensify water demand and strain inadequate infrastructure.
- *Poor water management:* Lack of integrated planning, inter-sectoral competition for water, and inefficient water use are common problems.

Continental Distribution of Freshwater

Continent	Share of Global Freshwater Resources (%)	Share of World Population (%)	Per Capita Water Availability (m^3/year)
Asia	32%	60%	3,000 – 4,500
Africa	9%	17%	2,000 – 3,500
Europe	8%	9%	5,000 – 9,000
North America	15%	5%	12,000 – 20,000
South America	26%	8%	23,000 – 40,000
Oceania	5%	1%	20,000+

Distribution of Surface and Groundwater by Continent

Surface Waters (Rivers and Lakes)

- *South America:* Home to the Amazon River and tropical rainforests; water resources are abundant.
- *Africa:* Includes major rivers such as the Nile and the Congo; however, water availability is highly uneven and regionally variable.
- *Asia:* Features large river systems like the Mekong, Ganges, and Yangtze; although water resources are significant, population density is extremely high.
- *Europe:* Major rivers include the Danube and the Volga; water availability is generally stable, though climate change is increasingly altering flow regimes.
- *North America:* The Mississippi River and the Great Lakes constitute well-managed and relatively well-preserved water systems.

Groundwater (Aquifers)

- *Asia:* The Ganges Basin aquifer in India is the most heavily utilized in the world.
- *Africa:* Sub-Saharan aquifers are vast but largely inaccessible due to depth and technical limitations.
- *North Africa and the Middle East:* Deep fossil groundwater reserves are being exploited—these are non-renewable water sources.
- *South America:* The Guarani aquifer is among the largest in the world.
- North America: The Ogallala aquifer, located beneath the Great Plains of the United States, is crucial for agriculture but is depleting rapidly. Due to this, water pricing mechanisms have been introduced to balance its use. For instance, irrigation from a 61-meter depth in Nebraska costs approximately $772 per hectare (Aydoğdu, 2012).

The global distribution of freshwater resources is highly unequal. Significant differences exist among continents, countries, and population densities in terms of per capita water availability. While a continent may appear to have sufficient water overall, national and regional disparities are often dramatic. For instance, South America benefits from the Amazon Basin, and Canada and Alaska have substantial reserves of groundwater and glacial water. In contrast, Asia (especially China and India) and Africa face severe water stress due to high population pressures, scarce resources, and inadequate infrastructure (Gleick, 2014). In Europe, Mediterranean countries experience seasonal and structural water stress. In Oceania, New Zealand is rich in water resources, while Australia is one of the world's driest continents.

Groundwater plays a critical role in regions where surface water is insufficient. However, over-extraction and pollution have led to its rapid depletion. Today, water is becoming such a strategic resource that it lies at the core of issues like climate change, migration, and conflicts (Aydoğdu, 2024).

Global, Continental, and National (Türkiye) Sectoral Water Use (% Distribution)

Region / Country	Agriculture (%)	Industry (%)	Domestic Use (%)	Source
World Average	69	19	12	FAO (2021); WWAP (2023)
Asia	81	11	8	FAO (2021)
Africa	88	5	7	WWAP (2023); UNEP (2022)
Europe	33	48	19	Eurostat (2022); WWAP (2023)
North America	37	41	22	USGS (2022); WWAP (2023)
South America	71	11	18	FAO (2021); CEPAL (2021)
Australia / Oceania	52	30	18	OECD (2023)
Türkiye	74	12	14	Ministry of Agriculture and Forestry (2023); TÜİK (2023)

Regional Variations in Sectoral Water Use

Differences in sectoral water use across regions are shaped by both economic activities and the structure of natural resources. On a global scale, agriculture accounts for the largest share of water use (FAO, 2021), underscoring its priority and critical role in water management. The countries with the highest agricultural water consumption are India, China, and the United States.

In developing regions such as Asia and Africa, agricultural water use represents 81% and 88%, respectively, directly linked to the prevalence of irrigation-dependent farming. Limited use of modern irrigation technologies in these regions exacerbates water inefficiency.

In contrast, in industrialized regions like Europe and North America, the share of industrial water use is significantly higher (48% and 41%, respectively). This reflects the high water demand of technological production processes and underscores the growing importance of wastewater management and treatment technologies.

In Türkiye, approximately 74% of total water consumption occurs in the agricultural sector, exceeding the global average. This indicates that agricultural irrigation represents the country's most significant risk area in water management. Particularly in southeastern regions (e.g., Harran Plain), traditional flood irrigation practices result in substantial water losses, deepening water stress.

Globally, urban (domestic) water use remains relatively low, but with population growth and urbanization, this share is expected to become increasingly critical in the coming decades. Rising temperatures and climate change will make drinking water supply more challenging. Moreover, access to water infrastructure is closely correlated with levels of socioeconomic development.

Country-Level Comparisons

Country	Per Capita Water (m^3/year)	Water Stress Status	Agricultural Use (%)
Canada	~80,000	None	12%
Brazil	~45,000	None	65%
China	~2,000	Moderate	65%
India	~1,400	High	85%
Egypt	~700	Critical	88%
Türkiye	~1,300	Water Stress	74%

Water scarcity is not solely determined by absolute water availability, but also by factors such as development level, consumption habits, infrastructure quality, and governance systems (Falkenmark & Widstrand, 1992; Küçük et al., 2025).

Türkiye's Water Assets and Utilization

A significant portion of Türkiye lies within a semi-arid climate zone. The country's total annual renewable freshwater potential is approximately 234 billion m³. However, not all of this water is technically or economically usable. Based on the amount of usable water per capita, Türkiye is classified among the "water-stressed countries."

- Average annual precipitation: 574 mm (below the global average)
- Renewable freshwater potential: ~112 billion m³
- Annual extractable groundwater: ~54 billion m³ (DSİ, 2023)
- Per capita annual available water: ~1,300 m³ → Water-stressed (According to UNDP classification: 1,000–1,700 m³ = stressed)

Regional Imbalances

- Highest water potential: Eastern Black Sea, Eastern Anatolia (except Iğdır and nearby areas)
- Lowest water potential: Central Anatolia (especially Konya, Karaman, Aksaray) and Southeastern Anatolia (particularly the Harran Plain)
- Most fragile water supply-demand balance: Konya, Karaman, Aksaray, Şanlıurfa, Iğdır

Major Issues

- Irrigation inefficiency: Around 80% of agricultural irrigation still relies on flood (surface) methods.
- Water losses: Leakage and unaccounted-for water in distribution networks can reach up to 40%.

- Groundwater depletion: Over-extraction is causing rapid declines in aquifer levels, resulting in sinkhole formation in some areas, a problem that continues to worsen.
- Increasing drought risk due to climate change.
- Unplanned urbanization and inadequate protection of water basins have become serious threats.

These problems indicate that not only water resources, but also soil, biodiversity, and the integrity of ecosystems are at risk. Therefore, environment, nature, and natural resources must be managed holistically and interdependently.

Türkiye–Global Comparison and Risk Analysis

Criterion	Global Average	Türkiye	Interpretation for Türkiye
Per capita water availability (m³)	~5,900	~1,300	Far below the global average
Agricultural water use	~70%	~74%	High usage despite stress conditions
Water losses	~25% (ideal)	40%+	Infrastructure investment urgently needed
Treatment capacity	~59%	47% (urban)	Industrial wastewater reuse insufficient
Water policy structure	Integrated in developed countries	Fragmented in Türkiye	Multiple responsible authorities; basin-based management not yet fully implemented

Global Sectoral Water Use and Efficiency (%)

Sector	Use Share (%)	Efficiency (Average)
Agriculture	69	40–60
Industry	19	70–90
Domestic (Municipal)	12	80–95

The growing global population and the resulting increase in food demand have made nearly all countries dependent on agricultural irrigation. Promoting advanced irrigation technologies that save water and increase productivity is a key factor in effective water demand management (Küçük et al., 2025).

Globally, surface irrigation remains dominant in the agricultural sector, leading to substantial water losses. Transitioning to pressurized irrigation systems and the reuse of treated wastewater is becoming an urgent necessity (Güneş et al., 2022). In the industrial sector, water efficiency improves through recycling, cooling systems, and closed-loop processes. These practices should be incentivized, regulated, and expanded. In domestic water use, efficiency is relatively high due to water metering, treatment, and greywater reuse, but such measures should be further promoted and widely implemented.

Sectoral Water Usage and Efficiency in Türkiye (%)

Sector	Share Usage (%)	Estimated Efficiency (%)
Agriculture	74	35–45
Industry	12	60–80
Domestic	14	75–85

Agricultural, Industrial, Domestic, and Energy-Based Water Use in Türkiye: Sectoral Analysis

In Türkiye, agricultural irrigation is predominantly carried out using traditional methods such as furrow irrigation, flood irrigation, basin irrigation, and wild irrigation. Modern pressurized irrigation systems (drip, sprinkler, pivot, etc.) are not yet widespread. The expansion of pressurized irrigation systems, along with increased incentives and financial support, is essential.

Water use in the industrial sector is gradually increasing. Although some organized industrial zones have adopted water recovery systems, such practices are not yet common nationwide. These systems should be made mandatory and more widely implemented.

Domestic water use is relatively high in urban areas; however, infrastructural deficiencies (leakages, old pipes, etc.) significantly reduce efficiency (Aydoğdu, 2023). Gradual renewal of water distribution networks and effective monitoring to prevent leaks and losses are necessary.

Sectoral Evaluation of the Water Problem

Water is of vital importance across all sectors, primarily agriculture, industry, and domestic use. However, each sector has distinct needs, consumption patterns, and impacts on water resources. Therefore, adopting a sectoral approach to water management is essential.

Agricultural Sector

Main General Problems

- Inefficient irrigation techniques: Surface irrigation methods (wild, flood, and furrow irrigation) are still widely used, resulting in low water conveyance and irrigation efficiency. Additionally, evaporation losses are high, leading to inefficient use of water.
- Unplanned cropping patterns: Water-intensive crops (e.g., maize, cotton, rice) are cultivated in hot and arid regions, i.e., in areas with water scarcity.
- Soil and water salinity: Excessive and improper irrigation practices disrupt soil structure and affect downstream water quality through natural flow, reducing productivity.
- Lack of adaptation to climate change: Agricultural systems are insufficiently prepared for extreme weather events such as frost and drought. Issues related to adaptation and resilience are widespread.

General Solution Proposals

- Expansion of pressurized irrigation systems such as drip, sprinkler, and pivot irrigation is mandatory.
- Crop planning should be based on water availability (i.e., products with a low water footprint).
- Farmers should be made more aware through education, extension, and support programs.
- Digital water management in agriculture (e.g., sensors, remote monitoring, and meteorological decision support systems) should be promoted.

Industrial Sector

Main General Problems

- Discharging untreated water into the environment contaminates surface and groundwater resources.
- Water-intensive industries: Sectors such as textiles, food processing, and chemicals consume large volumes of water.
- Insufficient treatment and recycling: The recovery and reuse of wastewater remain limited.
- Water use in energy production: Thermal power plants consume significant amounts of water.

General Solution Proposals

- Reuse (recycling) of water within industries should be expanded. Innovative water-saving methods in water-intensive sectors should be encouraged.
- Mandatory wastewater treatment and monitoring systems should be implemented, with strict enforcement.
- "Clean production" and "green industry" incentives should be increased.
- Water efficiency certification systems (such as ISO 14046 water footprint standards) should be emphasized.

Domestic and Urban Use

Main General Problems

- Infrastructure losses: Leakage and losses in distribution networks range between 30% and 40%.
- High individual consumption: Water use is excessive due to low public awareness (e.g., 150–250 liters per capita per day).
- Deficiencies in wastewater management: Some treatment facilities are inadequate or poorly operated.
- Drought risk: During dry periods, dam reservoir levels drop, increasing the likelihood of disruptions in urban water supply.

General Solution Proposals

- Urban water infrastructure must be renewed, and losses must be reduced.
- Greywater systems (e.g., reusing sink and shower water for garden irrigation) should be encouraged and expanded.
- Water-saving fixtures should be widely adopted in households and public buildings.
- Awareness campaigns (public education on water conservation, public service announcements, and reward mechanisms) should be effectively and sustainably implemented.

Energy Sector

Main General Problems

- Hydroelectric and thermal power plants, as well as cooling systems, consume large quantities of water.
- Especially thermal power plants both withdraw and discharge heated water back into the environment, impacting ecosystems and natural resources.
- Thermal pollution: The discharge of hot water into rivers, lakes, or surface waters alters thermal balance and disrupts aquatic life.
- Hydroelectric power plants (HEPPs) can alter natural flow regimes. Because turbines require continuous water flow to generate electricity, river regulation changes—particularly in climate-sensitive regions—can harm ecosystems.
- Drought reduces the production capacity of reservoir-based power plants.

General Solution Proposals

- Transitioning to water-efficient energy technologies is not an option but a necessity.
- Greater emphasis and incentives should be placed on renewable and non-water-dependent energy sources such as solar and wind.
- Existing energy facilities should enhance cooling water recovery and conduct more effective environmental impact assessments.

General Recommendations for Sectoral Water Use

- Water footprint accounting and transparent reporting: Every sector, product, and service must document and disclose water use data.
- Inter-sectoral coordination is essential in water management planning.
- Climate change adaptation policies must be effectively integrated into sectoral strategies.
- Public–private partnerships should be strengthened for R&D, technology transfer, and financing.
- Early warning systems for drought, frost, and extreme rainfall should be expanded.
- Climate-resilient infrastructure (dams, irrigation systems, and urban infrastructure) must be redesigned according to climate risks.
- Drought management plans: Provincial-level water crisis committees and strategies should be strengthened.
- Rainwater harvesting systems: Rain collection and storage technologies should be promoted through incentives and support programs.
- Decarbonization: Reducing fossil fuel use and promoting renewable energy are crucial to limiting global temperature rise.
- Pressurized and digital irrigation systems should be more strongly supported in agriculture.
- Efficient irrigation practices must be expanded through water pricing mechanisms and public incentives.
- Groundwater monitoring and usage restrictions (especially in arid regions) should be enforced.
- Water recycling in industry must be treated as a necessity.
- Urban infrastructure improvement and greywater reuse should be further encouraged and mainstreamed.

An Overview of Agricultural Irrigation Systems

Irrigation systems encompass the methods, infrastructure, and technologies used to transport water from its source to plants through specific technical and engineering practices when natural rainfall is insufficient for crop needs (FAO, 2021; Keller & Bliesner, 1990).

Irrigation is not only about conveying and distributing water but also about managing its timing, quantity, and efficiency. A well-designed irrigation system is of great importance for the conservation of water resources, enhancing agricultural productivity, and ensuring environmental and resource sustainability.

Classification of Irrigation Systems

Irrigation systems are generally classified into four main groups based on their water application methods:

1. *Surface Irrigation Systems*

Water is applied directly to the field and distributed across the soil surface by gravity to reach the plants. It is traditionally the most common method.

Subtypes:

- Furrow irrigation
- Basin (non-pressurized) irrigation
- Flood (wild) irrigation

Characteristics:

- One of the oldest and most widely known systems.
- Low investment cost but low water efficiency (approximately 40–60%).
- May be inefficient on sloped lands or soils with high permeability.

2. *Sprinkler Irrigation Systems*

Water is applied under pressure through pipes and sprinklers in a controlled manner, allowing water savings and better distribution.

Subtypes:

- Fixed systems
- Portable systems
- Center-pivot systems

Characteristics:

- Provide more uniform water distribution.
- Require moderate investment.
- Can be affected by wind and evaporation.
- Efficiency rates are typically 60–75%.

3. *Drip and Micro-Irrigation Systems*

Water is delivered drop by drop directly to the plant root zone under relatively low pressure. Drip irrigation is the primary method recommended for sustainable agriculture in regions facing water scarcity.

Characteristics:

- The highest water efficiency system (90–95%).
- Saves both water and fertilizer.
- Requires higher investment and maintenance costs compared to other systems.
- Widely used in greenhouses, orchards, and vegetable production.

4. *Subsurface Irrigation Systems*

Water is delivered directly to the root zone through underground pipes or channels.

Characteristics:

- Surface evaporation is minimal.
- Can be integrated with automation systems.
- Complex system structure with high initial investment cost.

Additional Classifications in Irrigation Systems

According to Water Source:

- Surface water (river, lake, stream, etc.)
- Groundwater (well)
- Reclaimed water (after wastewater treatment)

According to Control Type:

- Manual systems
- Semi-automatic systems
- Automatic systems (sensor-based, timer-controlled, IoT-supported)

Modern Trends and Sustainability Perspective

Considering climate change, increasing water demand, and polluted and diminishing resources, the choice of irrigation systems today must take into account not only economic but also ecological, social, and societal criteria.

Drip and sprinkler irrigation systems are of critical importance for the future of sustainable agriculture in terms of water efficiency. In agriculture-based economies such as Türkiye, the widespread adoption of these systems will be a fundamental component of water management reform.

Comparison of Water Use Efficiency in Irrigation Systems

Criterion	Flood (Wild) Irrigation	Sprinkler Irrigation	Drip Irrigation
Water Efficiency	35–40%	60–70%	85–95%
Water Consumption for Irrigated Crops (1 ha, m³/year)	~10,000–15,000	~6,000–8,000	~3,000–4,000
Water Loss Rate	Up to 50%	30–40%	5–15%

Comparison of Irrigation Investment Costs (Average Cost per Hectare, USD)

Cost Item	**Flood Irrigation**	**Sprinkler Irrigation**	**Drip Irrigation**
Installation Cost	Low (~200–500)	Medium (~1,000–1,500)	High (~1,500–2,000)
Energy Requirement	Low	Medium	High (pressure needed)
Maintenance Cost	Very Low	Medium	High (filtering, clog control)
Depreciation Period	None (no fixed system)	5–7 years	5–10 years

Government Support (Türkiye)	None or very limited	Up to 50% grant	50–70% grant-supported (TKDK, DAP, DOKAP, IPARD)

Although pressurized systems have higher initial investment costs, they provide a return on investment in the long term through savings in water and fertilizer use.

Additional Comparisons of Irrigation Systems

Criterion	Flood (Wild) Irrigation	Pressurized Irrigation (Sprinkler & Drip)
Soil Salinity Risk	High	Low
Erosion and Waterlogging Risk	High	Low
Fertilizer Efficiency	Low (nutrients washed away, limited root reach)	High (fertilizer can be delivered via irrigation)
Labor Requirement	High	Low
Suitability for Modern Agriculture	Not suitable	Suitable

Comparison of Advantages and Disadvantages of Irrigation Methods

Method	Advantages	Disadvantages
Flood (Wild) Irrigation	Low investment, simple application	Excessive water loss, low efficiency
Sprinkler Irrigation	Moderate cost, uniform distribution	Efficiency reduced under windy conditions
Drip Irrigation	Highest efficiency, compatible with fertigation	High investment, sensitive maintenance

Global Sectoral Water Efficiency and the Economic Value of Water

The agricultural sector consumes the largest share of water both globally and in Türkiye, yet its economic return remains significantly lower than that of other sectors. According to data collected from 86 countries representing 56% of the global population, the economic contribution of 1 m³ of water (in USD) across all sectors is estimated at $18.9 per m³ (UN-WATER, 2021).

Sector	Average Income ($/m³)	Explanation
Agriculture	0.6	Low productivity; for instance, irrigation water fees generally range between 0.02–0.08 USD/m³.
Industry	32.2	High added value; sectors such as energy, textiles, and chemicals exhibit strong water efficiency.
Services	112.2	The highest efficiency; in finance, tourism, and technology, water use is low while returns are high.

Sectoral Water Efficiency and the Economic Value of Water in Türkiye

Sector	Average Income ($/m³)	Explanation
Agriculture	0.10 – 0.50	Low efficiency; due to traditional irrigation methods and low-value crops, the economic value of water remains limited.
Industry	30 – 50	Medium to high efficiency; water use is managed effectively, particularly in organized industrial zones.
Services	80 – 100+	High efficiency; especially in tourism and finance sectors, water use is minimal while returns are substantial.

According to UN-WATER (2021), in Türkiye, the economic contribution of 1 m³ of water is calculated as $0.3/m³ in agriculture, $251.3/m³ in industry (including the value of hydropower generation in dams), and $84.1/m³ in services. When all water-consuming sectors are evaluated together, the average economic contribution of 1 m³ of water in Türkiye is approximately $14.1.

Natural Resources, Water, and the Ethics of Scarcity: Between Moderation, Responsibility, and Justice

The rapidly intensifying global water shortages and scarcities are no longer merely technical or economic issues—they have become a moral test for humanity. The depletion of resources now questions to what extent our relationship with nature, the environment, and natural assets is ethical, fair, and responsible. The ethics of scarcity offers a philosophical framework that deepens this inquiry.

"Scarcity is not merely a problem of resources—it is a test of virtue." (Sachs, 2015)

The ethics of scarcity is an ethical approach that defines the moral responsibilities of individuals, societies, and states regarding the allocation, consumption, and preservation of limited natural resources. Unlike the ethics of abundance, it rests on three fundamental principles (Jamieson, 2008):

Moderation: The question is not how much we can obtain, but how much we truly need to use.

Justice: How can scarce resources be distributed equitably among all parties?

Responsibility: Who is responsible for the current situation, and to what extent?

Water Scarcity: A Moral Tension Field

Water, while a divine gift bestowed through nature, is also a fundamental human right. Yet under conditions of scarcity, this right is not distributed equally—the powerful consume more, while vulnerable groups (such as women, children, the elderly, rural populations, and local communities) bear the greatest burden (UNDP, 2023).

Some Ethical Questions:

- Is access to water a privilege or a right?
- Does saving water for future generations require sacrificing today's comfort?
- Is luxury consumption (e.g., private pools, golf courses) ethical in times of scarcity?

Use of Water and Natural Resources within the Framework of Scarcity Ethics

Principle	Moral Interpretation	Practical Application
Justice	Fair distribution of resources	Water rights legislation
Virtue	Moderation, humility, contentment	Transition from surface irrigation to pressurized systems
Responsibility	Moral obligation toward future generations	Water footprint education and promotion of personal water-saving practices
Foresight	Long-term planning against ecological risks	Drought scenarios and emergency action plans

The Ethics of Scarcity in Beliefs and Philosophical Systems

Islam: In Islamic thought, moderation and the avoidance of waste are regarded as both moral and environmental imperatives. The use of water and other natural resources is guided by a principle of restraint — a recognition that human beings have the right to benefit from creation, yet this right is bound by gratitude, balance, and responsibility.

The sacred teachings of Islam emphasize the sanctity of water and strictly prohibit its waste. This perspective is not only a religious admonition but also an ecological principle of balance. Respecting water thus becomes an act of faith — a way of honoring both the Creator and the delicate order of nature entrusted to humanity's care.

Buddhism: The principle of Karuna (compassion) is based on the idea of not causing harm to any living being. Therefore, it advocates for the equal and compassionate sharing of resources, especially under conditions of scarcity.

Judaism: The Sabbatical Year (Shmita) acknowledges the rights of the land and nature. It grants natural resources a right to rest — forbidding cultivation, plowing, or grazing during the seventh year.

Taoism: Human intervention in nature leads to scarcity. "The best governance is that which aligns with the rhythm of nature." (Tao Te Ching)

In the face of scarcity, virtue lies not only in sharing but sometimes in renunciation itself (Naess, 1989).

Such teachings reflect a profound ecological consciousness long embedded within spiritual traditions.

Recommendations Based on the Ethics of Scarcity

- Water use should be prioritized as follows: Drinking water > Agriculture > Services > Industry > Luxury consumption.
- In water-scarce regions, restrictions should be placed on luxury uses (such as golf courses or private pools).
- The right to water for all individuals must be constitutionally guaranteed and supported by effective legislation.
- Local communities should be granted greater authority and participation in resource management.

Natural resources are not measured by how we behave in times of abundance, but by how moral we remain in times of scarcity. The ethics of scarcity proposes a relationship with nature and resources that is not only protective but also sharing-oriented, virtuous, and grounded in justice.

"Nature provides enough to satisfy every man's need, but not every man's greed." — Mahatma Gandhi

A Ten-Step Roadmap for Sustainable Water Management

1. *Transition to Basin-Based Integrated Water Management*

- Water management should be planned not according to political or administrative boundaries, but hydrological basins, based on the priorities of each basin and country.
- Inter-institutional data sharing, coordination, and joint action plans must become mandatory.

"Water knows no borders; therefore, planning must transcend them."

2. *Expansion of Pressurized and Smart Irrigation Systems in Agriculture*

- Support for pressurized irrigation systems should be increased through grants, incentives, subsidies, and credit systems accessible to all farmers, and their effectiveness should be regularly monitored.
- Smart sensor technologies, mobile monitoring platforms, and meteorological decision-support systems should be more actively used and expanded.

"Water saving = Food security + Climate resilience"

3. *Crop Planning Based on Water Availability*

- Crop patterns should be redesigned based on the water footprint approach, prioritizing water-sensitive, low-consumption, and substitute crops suited to each region.
- In arid and semi-arid regions, high-water-demand crops such as maize and cotton should be replaced with local, drought-resistant varieties and further incentivized.

"The soil does not choose, but water does."

4. *Modernization of Urban Water Infrastructure*

- Water loss and leakage rates in supply networks should be reduced from around 40% to below 25%, and ultimately to under 15%.
- Greywater systems (e.g., shower/sink water reuse for garden irrigation), rainwater harvesting, and storage systems should be standardized and encouraged through incentives and subsidies.

"Water efficiency must become an integral part of urban planning."

5. *Promotion of Water Recycling and Clean Production in Industry*

- Regulations on industrial water recovery, reuse, and mandatory treatment must be strengthened, with more rigorous inspections.
- Certifications such as ISO 14046 (Water Footprint) should be made compulsory, and green industrial zones should be more actively supported.

"Industry must advance—without increasing pollution or the total amount of water it consumes."

6. *Strict Monitoring and Protection of Groundwater Resources*

- Groundwater abstraction must be licensed, inspections intensified, and unauthorized wells permanently closed.
- Restriction zones and subsidence risk maps should be established; extraction quantities must remain in balance with recharge rates.

"What is out of sight must not be out of oversight."

7. *Establishment of Early Warning Systems for Droughts and Floods*

- Drought indices such as SPI, SPEI, and PDSI, combined with satellite data and climate projections, should be used to develop effective early warning centers.
- These systems should be expanded to provide SMS alerts directly to farmers.

"Management must begin before the crisis, not after."

8. *Legal Reforms Based on the Right to Water and the Ethics of Scarcity*

- Water should be defined not as a commodity, but as a human right, and its effective and fair use should be secured by law.
- Principles of equity, protection of vulnerable groups, and restriction of luxury consumption must guide water distribution.

"Access to water is not a privilege—it is a fundamental human right."

9. *Participatory Water Management: Involving All Segments of Society*

- Farmers, women, youth, industrial actors, and local NGOs must be included in decision-making processes.
- Basin commissions, irrigation associations/cooperatives, and municipal water councils should be strengthened and made more effective.

"Water democracy ensures water security."

10. *Water Education, Awareness, and Cultural Transformation*

- Water literacy should be incorporated into school curricula, and practical water education programs should be expanded, especially for farmers and water users.
- Media, mosques, and municipalities should help cultivate a culture of water conservation, promoting best practices through public campaigns.

"Manage water first with ethics and reason—then with the tap."

Multidisciplinary, Multi-Actor, and Long-Term Approach

Sustainable water management is not merely a technical or engineering process; it is also an ethical, ecological, and social transformation project.

The ten steps outlined above aim to leave future generations a livable water future through integrated institutional planning, sectoral sensitivity, and community-based participation.

Chapter Conclusion

Water is not merely a natural resource—it is the heart of nature, the test of humanity, and the key to the future. How we use water reflects our relationship with the environment, nature, and natural resources, while also serving as a test of our justice and moral values. Thus, sustainable water management must be approached not only through technical solutions but within a holistic framework encompassing ethical, cultural, ecological, and socio-political dimensions. In countries like Türkiye, located in semi-arid climatic zones, factors such as irrigation habits, infrastructural deficiencies, the impacts of climate change, and institutional fragmentation make water security a critical concern. When the low level of sectoral water efficiency is compared with the economic value of water, the situation becomes alarming. The ethics of scarcity proposed in this chapter offers a philosophical framework to redefine our attitude toward water—guiding us to leave behind a fairer, more virtuous, and responsible legacy for the future. Accordingly, all planning, investments, and policies must be built upon an ecologically balanced and justice-centered approach.

Chapter References

Altan, K., Teksoy, A., & Solmaz, S. K. A. (2020). *Türkiye'de Yağış ve Sıcaklığın Su Kaynakları, Tarımsal Ürün Verimi ve Su Politikalarına Etkisi (The Effect of Precipitation and Temperature on Water Resources, Agricultural Yield, and Water Policies in Turkey). Uludağ University Journal of The Faculty of Engineering*, 25(3), 1253–1270. https://doi.org/10.17482/uumfd.787493

Ateşoğlu, A., Aydoğdu, M. H., Yenigün, K., Sanchez-Paus Díaz, A., Marchi, G., & Bulut, F. S. (2025). *Insights from Earth Map: Unraveling Environmental Dynamics in the Euphrates–Tigris Basin. Sustainability*, 17, 7513. https://doi.org/10.3390/su17167513

Attfield, R. (1991). *The Ethics of Environmental Concern* (2nd ed.). University of Georgia Press. Retrieved from https://www.ugapress.org/9780820313443/the-ethics-of-environmental-concern/

Ayars, J. E., Phene, C. J., Hutmacher, R. B., Davis, K. R., Schoneman, R. A., Vail, S. S., & Mead, R. M. (1999). *Subsurface drip irrigation of row crops: A review of 15 years of research at the USDA-ARS, Water Management Research Laboratory. Agricultural Water Management*, 42(1), 1–27. https://doi.org/10.1016/S0378-3774(99)00025-6

Aydogdu, M. H., Karlı, B., Yenigün, K., & Aydogdu, M. (2015). *The farmers' views and expectations to the Water User Associations; GAP–Harran plain sampling, Turkey. Global Advanced Research Journal of Agricultural Science*, 4(1), 33–41.

Aydogdu, M. H., Karlı, B., Parlakcı Dogan, H., Sevinç, G., Eren, M. E., & Kücük, N. (2018). *Economic Analysis of Agricultural Water Usage Efficiency in the GAP-Harran Plain: Cotton Production Sampling, Sanlıurfa-Turkey. International Journal of Advances in Agriculture Sciences*, 3(12), 12–19.

Aydogdu, M. H., Karlı, B., Yenigün, K., & Aydogdu, M. (2016). *Evaluation of Farmers' Willingness to Pay for Water under Shortages: A case study of Harran Plain, Turkey. Journal of Environmental & Agricultural Sciences*, 7, 23–28.

Aydoğdu, M. H. (2012). *Şanlıurfa-Harran Ovasında Tarımda Su İşletmeciliği ve Fiyatlandırılması, Sorunlar ve Çözüm Önerileri (Water Management and Pricing in Agriculture in the Şanlıurfa-Harran Plain: Problems and Proposed Solutions)*

[Doctoral dissertation, Harran University]. Harran Üniversitesi Fen Bilimleri Enstitüsü, Şanlıurfa.

Aydoğdu, M. H. (2023). *Türkiye'de Son Çeyrek Yüzyılda Gerçekleşen Belediye İçme ve Kullanma Suyu Göstergelerinin Analizi (Analysis of Municipal Drinking and Utility Water Indicators in Turkey over the Last Quarter Century). Journal of Social, Humanities and Administrative Sciences*, 9(63), 2670–2675. https://doi.org/10.29228/JOSHAS.69196

Aydoğdu, M. H. (2024). *İklim Değişikliği: Sorunlar ve Bunlar Yönetilebilir mi? (Climate Change: Problems and Are They Manageable?) Social Science Development Journal*, 9(46), 79–91. https://doi.org/10.5281/zenodo.13765428

Aydoğdu, M. H. (2024). *İklim Değişikliğinin Sosyal ve Ekonomik Yapı Üzerine Etkilerinin Genel Değerlendirmesi (General Assessment of the Social and Economic Impacts of Climate Change).* In B. Güngör & Ö. Küçük (Eds.), *Sürdürülebilir Gelecek İçin İklim Perspektifleri (Climate Perspectives for a Sustainable Future)* (4th ed., pp. 65–82). Özgür Yayınları. https://doi.org/10.58830/ozgur.pub564.c2270

Aydoğdu, M. H., Sevinç, M. R., & Cançelik, M. (2021). *Determination of farmers' willingness to pay for drought adaptation policies in Şanlıurfa, Turkey. Weather, Climate, and Society*, 13, 677–686. https://doi.org/10.1175/WCAS-D-20-0163.1

Berry, T. (1999). *The Great Work: Our Way into the Future.* Bell Tower. Retrieved from
https://www.amazon.com/Great-Work-Our-into-Future/dp/0609804995

Burt, C., Clemmens, A., Strelkoff, T., Solomon, K., Bliesner, R., Hardy, L., & Howell, T. (1997). *Irrigation performance measures: Efficiency and uniformity. Journal of Irrigation and Drainage Engineering*, 123(6), 423–442. https://doi.org/10.1061/(ASCE)0733-9437(1997)123:6(423)

CEPAL. (2021). *Estadísticas del Agua en América Latina y el Caribe (Water Statistics in Latin America and the Caribbean).* Comisión Económica para América Latina y el Caribe. Retrieved from
https://www.cepal.org/es/temas/recursos-hidricos

Chapple, C. K. (1998). *Hinduism and Ecology: The Intersection of Earth, Sky, and Water.* Harvard University Press, UK.

Daly, H., & Farley, J. (2011). *Ecological Economics: Principles and Applications.* Island Press. Retrieved from https://library.uniteddiversity.coop/Measuring_Progress_and_Eco_Footprinting/Ecological_Economics-Principles_and_Applications.pdf

Denny, F. M. (1998). *Islam and ecology: A bestowed trust. Harvard Theological Review*, 91(2), 199–214.

Diyanet İşleri Başkanlığı. (2021). *Kur'an-ı Kerîm Meali (Translation of the Qur'an).* Retrieved from https://kuran.diyanet.gov.tr

DSİ. (2023). *Su kaynakları raporu (Water Resources Report).* Devlet Su İşleri Genel Müdürlüğü. Retrieved from https://www.dsi.gov.tr/

Eurostat. (2022). *Water Statistics.* Retrieved from https://ec.europa.eu/eurostat

Falkenmark, M., & Widstrand, C. (1992). *Population and water resources: A delicate balance. Population Bulletin*, 47(3), 1–36.

FAO. (2020). *The State of the World's Land and Water Resources for Food and Agriculture – Systems at breaking point.* Retrieved from https://www.fao.org/interactive/state-of-food-agriculture/2020/en/

FAO. (2021). *AQUASTAT Database.* Food and Agriculture Organization of the United Nations. Retrieved from https://www.fao.org/aquastat/

FAO. (2021). *Irrigation Manual for Training of Extension Staff and Farmers.* Food and Agriculture Organization of the United Nations. Retrieved from https://openknowledge.fao.org/server/api/core/bitstreams/37d3e821-2d61-4e72-bc85-0cdd9a9ad82e/content

Firidin, E. (2015). *Su Sorununun, Su Hakkı ve Su Etiği Çerçevesinde Değerlendirilmesi (Evaluation of the Water Problem within the Framework of Water Rights and Water Ethics). Aksaray Üniversitesi İktisadi ve İdari Bilimler Fakültesi Dergisi*, 7(2), 43–55.

Foltz, R. C. (2003). *Worldviews, religion, and the environment: A global anthology.* Wadsworth. Retrieved from https://archive.org/details/worldviewsreligi0000unse

Gleick, P. H. (2014). *The World's Water Volume 8: The Biennial Report on Freshwater Resources.* Retrieved from https://link.springer.com/book/10.5822/978-1-61091-483-3

Global Footprint Network. (2023). *Ecological Footprint data.* Retrieved from https://www.footprintnetwork.org

Gutas, D. (2001). *Avicenna and the Aristotelian tradition: Introduction to reading Avicenna's philosophical works.* Brill. Retrieved from https://books.google.com.tr/books/about/Avicenna_and_the_Aristotelian_Tradition.html?id=G10qnwEACAAJ&redir_esc=y

Güneş, K., Çullu, M. A., Şimşek, M., Topkaya, B., Aydoğdu, M. H., Beşiktaş, M., & Çelebi, D. M. (2022). *Chapter 6: Reuse of Water in Agriculture (Treated Wastewater, Drainage Water)* (pp. 231–245). In M. Bahadir & A. Haarstrick (Eds.), *Water and Wastewater Management* (Global Problems and Measures). Springer, Cham. https://doi.org/10.1007/978-3-030-95288-4_19

Haidi, A., Shakhavan, H. S., Dabbagh, O., Kadhim, D., & Çullu, M. A. (2024). *Comparing Urban Heat Islands in Erbil City-Iraq: Investigating Vegetation Response Through Day and Night Thermal Infrared Data And NDVI Values. Applied Ecology and Environmental Research*, 22, 1917–1930. https://doi.org/10.15666/aeer/2202_19171930

Jamieson, D. (2008). *Ethics and the Environment: An Introduction.* Cambridge University Press, UK.

Kabakcı, Y. (2025). *Hz. Muhammed Dönemi Medine'nin Su Kaynakları ve Onun Suya Dair Uygulamaları (Water Resources of Medina during the Prophet Muhammad Era and His Practices Related to Water). Divan Disiplinlerarası Çalışmalar Dergisi*, 30(59), 1–28. https://doi.org/10.20519/divan.1556786

Karataş, A. (2023). *Azalan ve Kirlenen Su Kaynaklarına Bir Önlem Olarak Çevre Eğitimi: Brezilya Örneği (Environmental Education as a Measure Against Decreasing and Polluted Water Resources: The Brazil Example). Sosyolojik Düşün*, 8(3). https://doi.org/10.37991/sosdus.1374371

Kaya, H., & Mazlum, H. (2025). *Hayvansal Gıda Üretiminde Su Ayak İzinin Değerlendirilmesi (Evaluation of Water Footprint in Animal Food Production).*

Osmaniye Korkut Ata Üniversitesi Fen Bilimleri Enstitüsü Dergisi, 8(2), 970–984. https://doi.org/10.47495/okufbed.1552393

Keller, J., & Bliesner, R. D. (1990). *Sprinkler and trickle irrigation.* Van Nostrand Reinhold, New York, US.

Kimence, T. (2023). *Su Kaynakları Yönetiminde Havza Ölçekli Faaliyetler (Basin-Scale Activities in Water Resources Management).* In *İklim Değişikliği Çerçevesinde Su Kaynaklarının Mevcut Durumu ve Geleceği.* https://doi.org/10.53478/TUBA.978-625-8352-56-6.ch08

Kocabaş, H. M., Mut, R., Karakelle, E., & Kivilcim, M. (2025). *İklim Değişikliğine Uyum Bağlamında Su Yönetiminde Güncel Yaklaşımlar: Hollanda, İspanya ve Çin Halk Cumhuriyeti Örneklerinin İncelenmesi (Current Approaches in Water Management under Climate Change Adaptation: Case Studies from Netherlands, Spain, and China). JENAS Journal of Environmental and Natural Studies*, 7(2), 142–169. https://doi.org/10.53472/jenas.1743135

Koyuncu, Z., & Batdal, B. (2022). *İklim Değişikliğinin Su Kaynaklarına Etkisi (Impact of Climate Change on Water Resources).* Retrieved from https://www.researchgate.net/publication/394738761_Iklim_Degisikliginin_Su_Kaynaklarina_Etkisi

Küçük, N., Aydoğdu, M. H., Sevinç, M. R., Parlakçı Doğan, H., & Yenigün, İ. (2025). *Analysis of the Factors Determining the Agricultural Enterprises' Following Irrigation Technologies: Case Study of Şanlıurfa, Türkiye. Environmental Engineering and Management Journal*, 24(5), 967–977. http://doi.org/10.30638/eemj.2025.073

Maden, T. E. (2023). *Irak'ta Su Kaynakları Yönetimi (Water Resources Management in Iraq).* ORSAM Rapor No: 122, ORSAM Su Araştırmaları Programı Rapor No: 15. ISBN: 978-605-4615-17-9

McKee, T. B., Doesken, N. J., & Kleist, J. (1993). *The relationship of drought frequency and duration to time scales.* Proceedings of the 8th Conference on Applied Climatology, 17(22), 179–183.

Naess, A. (1989). *Ecology, Community and Lifestyle: Outline of an Ecosophy.* Cambridge University Press, UK.

Nasr, S. H. (1996). *Religion and the Order of Nature.* Oxford University Press, UK

Nasr, S. H. (2003). *The heart of Islam: Enduring values for humanity.* HarperOne. Retrieved from https://www.amazon.com/Heart-Islam-Enduring-Values-Humanity/dp/0060730641

National Drought Mitigation Center (NDMC). (2023). *U.S. Drought Monitor.* Retrieved from https://droughtmonitor.unl.edu/

Odum, E. P., & Barrett, G. W. (2005). *Fundamentals of Ecology* (5th ed.). Brooks/Cole. Retrieved from https://books.google.com.tr/books/about/Fundamentals_of_Ecology.html?id=vC9FAQAAIAAJ&redir_esc=y

OECD. (2012). *Water Governance in OECD Countries: A Multi-level Approach.* OECD Publishing. https://doi.org/10.1787/9789264174540-en

OECD. (2023). *Water Governance in OECD Countries: Policy Highlights.* Retrieved from https://www.oecd.org/en/publications/water-governance-in-oecd-countries_9789264119284-en.html

Özçalık, M., & Cobanoglu, O. (2023). *Karız Su Sisteminin Uygur Kültüründeki Sosyokültürel Değerinin Araştırılması* (Investigation of the Socio-Cultural Value of the Karez Water System in Uyghur Culture). *Uluslararası Uygur Araştırmaları Dergisi*, 22. https://doi.org/10.46400/uygur.1321782

Sachs, J. D. (2015). *The Age of Sustainable Development.* Columbia University Press, US.

Shiklomanov, I. A. (1993). World fresh water resources. In P. H. Gleick (Ed.), *Water in Crisis* (pp. 13–24). Oxford University Press, UK.

Shiva, V. (2002). *Water Wars: Privatization, Pollution, and Profit.* South End Press. Retrieved from https://archive.org/details/waterwarsprivati0000shiv

Strang, V. (2004). *The Meaning of Water.* Berg Publishers. https://doi.org/10.4324/9781003087090

Şıltu, E., & Akca, L. (2023). *İklim Değişikliği, Türkiye Yeraltı ve Yüzeysel Su Kaynakları* (Climate Change, Turkey's Groundwater and Surface Water

Resources). In *İklim Değişikliği Çerçevesinde Su Kaynaklarının Mevcut Durumu ve Geleceği.* https://doi.org/10.53478/TUBA.978-625-8352-56-6.ch01

T.C. Tarım ve Orman Bakanlığı. (2023). *Türkiye Tarımsal Kuraklık Bilgi Sistemi (TARBİS)* (Turkey Agricultural Drought Information System). Retrieved from https://www.tarimorman.gov.tr/

T.C. Tarım ve Orman Bakanlığı. (2023). *Ulusal Su Planı* (National Water Plan). Retrieved from https://www.tarimorman.gov.tr/SYGM/Belgeler/NHYP%20DEN%C4%B0Z/ULUSAL%20SU%20PLANI.pdf

Turhan, T., Paka, B. A. H., Wulandari, T. J., & Mustafa Şeker, M. (2025). *İslam ve Çevre Yönetimi: Sürdürülebilir Su Ekolojisine Bütüncül Bir Yaklaşım* (Islam and Environmental Management: A Holistic Approach to Sustainable Water Ecology). *Kent Akademisi*, 18(4), 1961–1987. https://doi.org/10.35674/kent.1624594

TÜİK. (2023). *Su Kaynaklarının Sektörel Kullanımı İstatistikleri* (Statistics on Sectoral Use of Water Resources). Türkiye İstatistik Kurumu. Retrieved from https://www.tuik.gov.tr/

UNDP. (2023). *Water scarcity in the MENA region and Turkey: Policy implications.* Retrieved from https://www.undp.org/water

UNEP. (2022). *Africa Water Atlas.* United Nations Environment Programme. Retrieved from https://www.unep.org/

UNESCO-WWAP. (2023). *United Nations World Water Development Report 2023: Partnerships and Cooperation for Water.* United Nations Educational, Scientific and Cultural Organization. Retrieved from https://www.unwater.org/publications/un-world-water-development-report

UNICEF & WHO. (2022). *Progress on Household Drinking Water, Sanitation and Hygiene 2000–2022.* World Health Organization (WHO) and the United Nations Children's Fund (UNICEF). Retrieved from https://www.who.int/publications/m/item/progress-on-household-drinking-water--sanitation-and-hygiene-2000-2022---special-focus-on-gender

UN-WATER. (2021). *Progress on change in water-use efficiency: Global Status and Acceleration Needs for SDG Indicator 6.4.1.* Retrieved from https://www.unwater.org/sites/default/files/app/uploads/2021/08/SDG6_Indicator_Report_641_Progress-on-Water-Use-Efficiency_2021_ENGLISH_pages-1.pdf

USGS. (2022). *Water Use in the United States.* United States Geological Survey. Retrieved from https://www.usgs.gov/mission-areas/water-resources/science/water-use-united-states

Vicente-Serrano, S. M., Beguería, S., & López-Moreno, J. I. (2010). *A multiscalar drought index sensitive to global warming: The Standardized Precipitation Evapotranspiration Index. Journal of Climate, 23*(7), 1696–1718. https://doi.org/10.1175/2009JCLI2909.1

WWAP (World Water Assessment Programme). (2023). *The United Nations World Water Development Report 2023: Partnerships and Cooperation for Water.* UNESCO. Retrieved from https://www.unesco.org/en/wwap

Yılmaz, F., & Sever, A. (2025). *Türkiye'de Kuraklık Riski ve Su Verimliliği* (Drought Risk and Water Efficiency in Turkey). Retrieved from https://www.researchgate.net/publication/394281834_Turkiye'de_Kuraklik_Riski_ve_Su_Verimliligi

Yılmaz, M. T. (2025). *İklim Değişikliğinin Afetler Üzerine Etkisi* (The Impact of Climate Change on Disasters). *ÇEM Hidrometeorolojik Afetler Eğitimi, Kastamonu.*

5th CHAPTER

EROSION AND NATURE: THE ANATOMY OF A SILENT EXHAUSTION

Chapter Summary

This chapter addresses the multifaceted impacts of erosion within the context of the soil's vital role in nature and the interaction between humans, nature, and natural resources through a holistic lens. Erosion is defined not merely as a physical process but as a multidimensional crisis with socioeconomic, cultural, philosophical, and religious dimensions. The study examines the types, causes, and ecological and social effects of erosion, while offering a vision for sustainable soil management grounded in Islamic environmental ethics, traditional knowledge systems, and nature-based solutions. Supported by data from the Turkish case, the analysis sheds light on the regional distribution, economic costs, and policy weaknesses related to erosion. Ultimately, it emphasizes that erosion is not only an environmental issue but also a matter of justice for humanity and future generations.

***Keywords*:** Soil Erosion, Sustainable Land Management, Islamic Environmental Ethics, Socioeconomic Impacts, Soil Policies in Türkiye

The Silent Cry of the Soil

Soil is one of the most vital components of nature and natural resources. It is not merely a physical surface but also a carrier of biodiversity, a filter of water, a regulator of climate, and the very foundation of civilizations. However, the exploitative relationship that humankind has established with nature and its resources increasingly undermines the soil's ability to fulfill these essential functions. In this context, erosion serves as a warning from nature itself. The loss of soil is not only an environmental issue but also an ethical and existential problem.

"Every civilization that loses its soil eventually loses itself." (Montgomery, 2007)

Erosion refers to the process of detachment, transport, and deposition of soil particles, leading to complex ecological, geomorphological, social, and economic consequences. This process may occur due to natural factors or as a result of human activities. The root causes of erosion lie in the reduction of vegetation cover that protects the soil's structure and integrity, its physical fragmentation, and changing environmental conditions (Lal, 2001). Erosion results not only in the physical loss of soil but also in the decline of soil fertility, deterioration of water quality, reduction of carbon storage capacity, and loss of biodiversity (Montgomery, 2007).

The Meaning of Erosion in the Human–Nature Relationship

Erosion is among the strongest indicators of the problematic relationship humanity has established with nature and natural resources. The "modern" paradigm of production and growth-oriented development has rendered natural assets subject to exploitation. Humanity has long exhibited a tendency to dominate rather than coexist with nature. Agricultural systems based on consumption rather than stewardship have deepened this alienation. Heidegger's critique of "reducing nature to an object" provides a solid philosophical foundation for understanding this rupture (Heidegger, 1977).

Islamic, Philosophical, and Cultural Perspectives: The Memory of the Soil

Soil is not only an ecological entity but also a cultural one. Many civilizations have referred to the soil as "mother," "father," or "ancestor," indicating its sacred place in human existence. Soil possesses a form of memory—it holds the past within itself. Archaeological remnants, microorganisms, and seed banks are fragments of this collective memory. Erosion erases this memory, and such a loss can transform into a loss of cultural identity (Nabhan, 1997).

In eco-philosophical thought, approaches such as deep ecology and eco-phenomenology emphasize the intrinsic value of nature. According to these views, soil is not merely a possession of humankind but a living component that holds rights of its own (Naess, 1973). In local cultures, respect for the soil is deeply rooted. In Anatolia, cultural practices such as "praying before stepping on the soil" or "kissing the earth" reflect traces of this ancient reverence. The loss of such connections in "modern" society has transformed erosion from a physical degradation into a spiritual disconnection. When the soil disappears, life largely disappears with it. Erosion is thus the most tangible sign of the breakdown in the human–nature relationship. The collapse of ecosystem services, threats to food security and safety, and the erosion of cultural memory are among its inevitable consequences. We must learn to perceive soil not merely as a "resource," but as a shared living space, an ethical responsibility, and a living being.

The Significance of Soil in Islam: A Theological, Ethical, and Legal Perspective

From a theological perspective, soil represents the source of both creation and fertility. In Islamic belief, the soil is the origin of life and a manifestation of God's creative power. In many spiritual traditions, the earth is regarded as the origin of creation and the source of abundance. Within Islamic thought, it represents both the foundation of life and a reflection of divine power. The Qur'an frequently reminds humanity of its deep connection to the soil — as the material from which life was formed, the resting place to which all return, and the ground from which life will rise again.

In this sense, the earth embodies not only a biological reality but also a metaphysical truth — a symbol of continuity, renewal, and divine wisdom. It is not merely matter; it is a living sign of the sacred order that sustains all existence, nurturing every form of life and bearing witness to creation itself.

Prophetic Heritage: A Life in Harmony with the Soil

In Islamic thought and cultural tradition, soil is not merely the material foundation of life, but also a symbol of purification, humility, and renewal. In the life of Prophet Muhammad (peace be upon him), one observes a profound respect and sensitivity toward the earth. He regarded soil as a means of both physical and spiritual cleansing. The hadith, "The earth has been made for me a place of prostration and a means of purification" (Bukhari, Tayammum 1; Muslim, Masajid 3), reflects this understanding. It shows that the human connection with nature is not only biological but also devotional — a form of worship through unity with creation.

In this sense, soil represents one of the most silent yet deepest bonds between humankind and the Creator. When water is unavailable, the act of tayammum — symbolic purification with earth — signifies a spiritual return to one's origin and a reaffirmation of divine order. The Prophet's saying, "Even if the Day of Judgment arrives and you have a sapling in your hand, plant it," deepens this vision. It reflects not merely environmental sensitivity but also faith, responsibility, and trust in the eternal cycle of life.

This spiritual understanding finds rich expression in Sufi thought. Yunus Emre, in his poem "The Yellow Flower," conveys the cyclical nature of existence and the human return to essence through simple yet profound symbolism. When he asks the flower about its parents, the flower answers that they are "the earth itself." Through this dialogue, Yunus reveals a timeless truth — that humanity originates from the soil and ultimately returns to it, attaining divine unity within this sacred cycle. Thus, in Yunus's vision, the earth is not the end of life but its eternal beginning.

Similarly, Jalāl al-Dīn Rūmī expresses this metaphysical relationship through his famous counsel: "Be humble and self-effacing, like the earth." For Rūmī, the soil embodies both being and non-being, acceptance and transformation. The earth bears all things — the pure and the impure, the beautiful and the broken — and through patience and transformation, it returns them as life, beauty, and abundance. Thus, soil becomes the quiet teacher of creation and renewal.

Humility, in this light, is not a passive surrender but a conscious participation in the rhythm of existence — an act of aligning oneself with the law of transformation. The humble person, like the earth, turns harm into healing, darkness into light, and through gentleness gives life to all around. In this way, the soil becomes the mirror of human spiritual maturity.

Therefore, in the Qur'an, as well as in the words of Yunus Emre and Rūmī, the soil is not simply a physical substance, but a living metaphor for humanity's place in creation, its power to transform, and its participation in divine harmony. The earth is both the home of the body and the teacher of the soul — a sacred reminder that every beginning and every return share the same essence: life emerging endlessly from the dust, and the dust returning eternally to life.

Islamic Environmental Ethics and Soil Rights

In Islamic thought, the earth is regarded not merely as a physical domain for life, but as a sacred trust in which humanity fulfills its divine responsibility. According to the Qur'an, the ultimate ownership of the earth belongs to God, while human beings dwell upon it as trustees and stewards (khalifah). This perspective defines humanity's role not in terms of possession and domination, but through accountability, care, and protection. Thus, the land is not an object of exploitation, but a sacred entity entrusted to human conscience.

The legal and moral principles of Islam frame this relationship within a balance of justice and sustainability. Cultivating the land and making it productive is seen as a form of service to both creation and community. This balance between environmental stewardship and social welfare is reflected in classical Islamic jurisprudence through principles such as ihyā' al-mawāt — the revival of barren or neglected lands through cultivation and care. In this way, the human role is not only to use the soil, but to revive, nurture, and sustain it.

In Islam, the connection to the earth also extends beyond life — it encompasses death and worship. Burial directly in the soil symbolizes both humility and the return to one's origin, affirming the cyclical nature of existence. Likewise, the act of prostration (sujūd) during prayer, in which the believer touches the ground with their forehead, signifies submission, reverence, and closeness to the Creator. The Qur'anic verse, "Prostrate and draw near" (Al-'Alaq, 96:19), expresses this sacred intimacy between human and earth as an act of devotion.

Thus, within the Islamic worldview, the earth is not simply the ground beneath human feet, but a space of trust, humility, and spiritual reflection. Humanity is not the ruler of the earth, but its caretaker — a being who lives with the land, protects it, and ensures its continuity. This holistic vision regards nature as both a sign of divine order and a field in which moral responsibility is tested.

The Ethical and Ecological Dimensions of Soil

From a broader ethical standpoint, soil represents the intersection of ecology, economy, and morality. It provides food, supports ecosystems, and sustains human civilization—yet its degradation reflects the moral failures of human societies. When erosion strips the land, it simultaneously erodes social stability, cultural identity, and intergenerational equity.

Soil conservation, therefore, is not a merely technical or agricultural task; it is a moral, spiritual, and civilizational duty. It embodies the principle that caring for the earth is inseparable from caring for human dignity. In this sense, erosion is both a physical and metaphysical loss—an indicator of the imbalance between consumption and compassion, between human ambition and natural limits.

Erosion as an Ethical and Justice Issue

Erosion, in its deepest essence, is a question of justice. The loss of soil deprives not only the present generation but also the unborn of their right to sustenance and ecological security. It widens inequalities—between regions, between communities, and between the human and non-human worlds.

To confront erosion, societies must transcend short-term economic logic and adopt a moral ecology—an approach that unites ethical responsibility with environmental stewardship.

As such, the fight against erosion becomes a fight for justice, memory, and existence itself. The soil that nourishes us is also the ground of our moral being; how we treat it reflects who we are as a civilization.

Islamic Soil Ethics and Environmental Responsibility in the Modern Age

In the face of growing environmental challenges, the Islamic teachings concerning the earth and nature have regained significance. The ecological crises of the modern age reveal a deep rupture in humanity's sacred relationship with nature. According to Islamic thinkers, this separation is not merely environmental but also spiritual — a reflection of humankind's alienation from the divine order of creation. Thus, Islam's holistic worldview calls humanity to restore harmony (mīzān) with nature and to rediscover the balance embedded in creation.

The Qur'an declares: "God raised the heaven and established the balance; do not disturb that balance." (Ar-Rahman, 55:7–8). This balance extends beyond human relations — it encompasses the connection between humans and the natural world, especially the soil. In Islam, the earth is not viewed solely as a means of production or an economic resource, but as a sacred entity — the ground of life and death, the place of worship, and a mirror of divine wisdom.

From this perspective, protecting the soil and the environment is not only an ecological duty but also a moral and spiritual obligation. Qur'anic verses, prophetic traditions, and the principles of Islamic jurisprudence collectively emphasize the avoidance of waste, the preservation of natural balance, and respect for all forms of creation. This approach offers an ethical framework rooted in faith — one that invites humanity to coexist with nature, not as its master, but as its responsible guardian. This understanding provides a spiritual and ethical framework for addressing today's ecological challenges from an Islamic perspective.

To stand upon the earth is, in essence, to remember our origin and our return. The soil beneath our feet is not inert matter — it is the silent witness of creation, the keeper of memory, and the promise of renewal. In Islamic thought, humility before the earth is humility before the Creator; to harm it is to wound the very trust (amānah) that defines human existence. Modern civilization's estrangement from this sacred trust has turned the soil — once a symbol of life — into an object of exploitation. Reclaiming reverence for the earth, therefore, is not merely an ecological necessity but a spiritual restoration. It is a return to the balance where faith, ethics, and existence meet — a reminder that to care for the earth is, ultimately, to care for ourselves.

Definition and Types of Erosion

The process of erosion is generally divided into two main categories:

1. *Natural (Geological) Erosion*

Natural erosion encompasses the long-term processes of soil and rock weathering and displacement that occur over millions of years, shaping the Earth's surface through geological transformation. This type of erosion is considered a component of ecological equilibrium and typically progresses slowly. In environments where vegetation cover, soil depth, and climatic conditions are in harmony, there exists a natural balance between soil formation and loss (Morgan, 2005).

In essence, erosion can be viewed not only as a physical process of soil removal but also as a manifestation of the divine order inherent in creation. From a cosmic perspective, after the Big Bang, the Earth—initially a solid, lifeless mass formed through rapid cooling—underwent geological, chemical, and biological transformations that eventually rendered it habitable and agriculturally fertile.

Thus, erosion has not merely been a destructive force but an instrument of creation, shaping a "life-bearing surface" through the long evolutionary course of existence. The Earth's present form is the outcome of billions of years of natural weathering and reconstruction cycles. From this perspective, the "gentle face" of erosion lies in its role in the birth of soil—the very foundation of life.

Bediüzzaman Said Nursî, in his Isharat al-I'jaz, interprets the creation of the world within a metaphysical framework. He perceives the transformation of the Earth as the gradual manifestation of a "wise and purposeful divine power." In this sense, soil is not merely a physical substance but a realm that symbolizes the continuity of creation itself. Therefore, while erosion may appear as "loss" through the human eye, in the cosmic dimension it signifies renewal and rebirth—a perpetual act of divine recreation.

2. *Accelerated (Anthropogenic) Erosion*

Accelerated erosion occurs when human activities disturb natural equilibrium, causing soil loss to exceed the rate of soil formation. Improper land use in agriculture, deforestation, overgrazing, unplanned construction on slopes, and infrastructure development are among the primary triggers (Pimentel & Burgess, 2013).

Accelerated erosion substantially reduces soil fertility and water-holding capacity, threatening not only agricultural productivity but also food security and ecological stability (Panagos et al., 2015).

Erosion types are generally categorized as water erosion, wind erosion, gravitational erosion, and anthropogenic erosion.

Water Erosion

Water erosion refers to the detachment and transportation of soil particles through rainfall and surface runoff. It is typically classified into four subtypes:

- *Splash Erosion:* Caused by the impact of raindrops that dislodge and scatter loose soil particles, breaking down the surface structure (Lal, 2001).
- *Sheet Erosion:* The uniform removal of thin soil layers by surface flow, without forming visible channels.
- *Rill Erosion:* The formation of small, shallow channels on the soil surface as runoff concentrates.
- *Gully Erosion:* The deepening and widening of channels over time, which may render agricultural lands unusable (Morgan, 2005).

Wind Erosion

Wind erosion occurs primarily in semi-arid and arid regions where sparse vegetation cover leaves soil particles vulnerable to transport by wind. It unfolds in three stages: saltation, suspension, and surface creep (Toy et al., 2002).

Wind erosion not only alters the physical structure of soil but also degrades air quality and poses health hazards through the dispersion of fine dust particles (Goudie, 2014).

Gravitational Erosion

Gravitational erosion encompasses soil movements driven by the force of gravity, including landslides, rock falls, and soil slips. This form of erosion commonly occurs in steep terrains, saturated soils, or areas lacking vegetation cover. It can lead to sudden disasters, causing both loss of life and property (Selby, 1993).

Landslides: A Distinct Category among Erosion Types

Among the various types of erosion, landslides hold particular importance due to their destructive potential and frequency. Some major landslide events observed in Türkiye and worldwide are listed below (Froude & Petley, 2018; Petley, 2024; Wikipedia, 2024a; Wikipedia, 2024b; Öztürk et al., 2020).

Scope	Number of Events & Deaths	Notable Locations/Events
World (2004–2016)	~4,862 events, ~56,000 deaths	Widespread global distribution
World (2024)	766 fatal events, 4,933 deaths	Wayanad (India) and Enga (Papua New Guinea)
Future Projection (SSP2-4.5)	Annual deaths projected to increase from ~3,240 → ~8,380	Expected rise particularly in Asia and Africa
Türkiye (Annual Average)	4 fatal events, ~15 deaths	Concentration in Eastern Black Sea Region
Türkiye (1929–2019)	389 events, 1,343 deaths	Trabzon–Rize, Senirkent (Isparta)
High-Activity Period (2020s)	322 settlements, 2,000+ landslides	Ordu, Samsun, Trabzon – triggered by snowmelt

Anthropogenic Erosion

Anthropogenic erosion refers to soil degradation directly caused by human activities, often overlapping with accelerated erosion processes. Practices such as monoculture agriculture, stubble burning, deforestation, overgrazing, mining, road construction, and urbanization remove protective soil cover, disrupting natural balance and accelerating erosion rates.

This form of erosion contributes to the deterioration of ecosystem services, sediment pollution in water bodies, and disturbances in the global carbon cycle (Hooke, 2000).

Erosion is not merely a physical loss of soil; it is a multidimensional process that affects socio-economic systems, ecological functioning, and climate regulation. The accurate identification of erosion types is crucial for developing effective mitigation strategies. Therefore, both natural and anthropogenic causes must be jointly evaluated to strengthen sustainable land management practices.

Causes of Erosion

The causes of erosion are shaped by the complex interaction between natural processes and human activities. While the climatic, geological, and biological characteristics of a region determine the natural dynamics of erosion, anthropogenic interventions can significantly accelerate this process—often leading to irreversible ecosystem degradation (Lal, 2001; Montgomery, 2007). Below, these causes are examined under two main categories: natural and anthropogenic (human-induced).

Natural Causes

Natural drivers of erosion include climatic factors, topography (slope and landform), soil structure and properties, and lack of vegetation cover.

Climatic Factors

Climate is one of the most decisive elements influencing erosion. Intense and sudden rainfall events exceed the soil's infiltration capacity, accelerating surface runoff and detachment. In tropical regions, frequent and high-intensity precipitation results in dominant water erosion, whereas in arid and semi-arid zones, wind erosion prevails (FAO, 2021). After prolonged droughts, sudden heavy rainfalls can have the most devastating impacts, as the dry, unprotected soil structure becomes easily erodible. With climate change disrupting global precipitation regimes, erosion risks are increasing worldwide (IPCC, 2019).

Topography (Slope and Landform)

On steep slopes, gravitational forces intensify surface runoff. When the slope exceeds 15%, the soil's transport capacity increases significantly (Morgan, 2005). This poses a greater threat in mountainous agricultural areas. Furthermore, convex (outward-curved) and concave (inward-curved) landforms, located between valleys and ridges, exhibit varying soil movement dynamics. These micro-topographical characteristics directly influence both the type and severity of erosion.

Soil Structure and Properties

The texture, aggregation, and organic matter content of soil determine its susceptibility to erosion. Sandy soils allow rapid infiltration but are easily transported due to weak cohesion. In contrast, clay-rich soils retain water effectively but, when saturated, become sticky and prone to land sliding (Lal, 2001). Organic matter-rich soils are more resistant to erosion due to their enhanced water retention and aggregate stability (Pimentel & Burgess, 2013).

Lack of Vegetation Cover

Natural vegetation constitutes the first and most effective defense line against erosion. Plants stabilize soil through their root systems, protect it from raindrop impact, and reduce evaporation, maintaining soil moisture balance. The absence of vegetation—especially during rainy seasons—significantly increases surface runoff and soil displacement (Toy et al., 2002).

Anthropogenic Causes

Human-induced causes include improper agricultural practices, deforestation, overgrazing, industrialization, and urban expansion.

Improper Agricultural Practices

Agriculture is one of the most significant anthropogenic drivers accelerating erosion. Plowing perpendicular to slope weakens soil structure, obstructs runoff regulation, and facilitates soil displacement. Monoculture cropping disrupts the nutrient cycle and diminishes biodiversity, while excessive use of chemical fertilizers and pesticides weakens microbial activity and breaks down soil aggregates, reducing structural stability (Pimentel & Burgess, 2013; Lal, 2015).

Deforestation

Forests are the most effective natural mechanism for erosion control, especially in sloped terrains. In Türkiye, approximately 3 million hectares of forest land have been degraded since 1950 (TEMA, 2023). As a result of deforestation, the exposure of the soil surface increases both surface runoff and wind erosion, while also elevating carbon emissions, thereby negatively affecting the climate system (Hooke, 2000).

Overgrazing

Excessive and uncontrolled grazing by livestock, particularly in semi-arid regions, leads to the weakening of vegetation cover and the exposure of bare soil surfaces. This condition accelerates surface runoff and releases soil particles that can be easily transported by wind. According to FAO (2021), 35% of the world's rangelands are degraded, and the primary cause of this degradation is overgrazing.

Industrialization and Urban Expansion

Industrialization and urbanization processes have replaced natural landscapes with impervious surfaces such as asphalt and concrete. This prevents rainwater from infiltrating into the soil, transforming it into direct surface runoff, thereby increasing both erosion and flooding risks. Furthermore, excavation and storage of soil during construction activities disrupt the structural integrity of soil layers, making them more vulnerable to wind and water erosion (Montgomery, 2007).

While erosion caused by natural factors typically occurs over long periods within the balance of ecosystems, accelerated erosion triggered by human activities can cause extensive damage over a very short time. According to FAO (2021), in regions where natural erosion results in losses of only a few tons per hectare per year, improper land use can elevate these losses up to 100 tons per hectare per year. This comparison highlights the protective role of natural systems and the destructive potential of human influence.

Ecological Impacts of Erosion

Erosion is not merely the physical loss of soil; it is a multidimensional environmental problem that directly threatens ecosystem integrity, biological cycles, and environmental sustainability. The physical displacement of soil also removes nutrients, microorganisms, and biodiversity, thereby disturbing the natural balance and negatively affecting ecosystem services (Montgomery, 2007; Lal, 2004). The main ecological impacts of erosion are outlined below.

Decline in Soil Fertility

The most direct and widespread impact of erosion is the loss of the topsoil layer (A horizon), which supports agricultural productivity. This layer contains the highest concentrations of plant nutrients, organic matter, and microorganisms (Lal, 2001). The removal of this layer reduces the soil's water-holding capacity, aeration, and biological activity. The loss of just 2–3 cm of topsoil can cause yield reductions of up to 50% (Pimentel & Burgess, 2013).

According to FAO (2022), 33% of agricultural soils worldwide are moderately or severely degraded. In sub-Saharan Africa, soil losses average about 30 tons per hectare per year, directly reducing productivity (Montgomery, 2007). By contrast, in some regions of Europe where soil conservation measures are implemented, soil loss is limited to 1–2 tons per hectare per year (Panagos et al., 2015). This comparison clearly demonstrates the effectiveness of sustainable land management policies.

Decline in Water Quality and Sedimentation

Fine soil particles transported by erosion enter surface waters, ponds, and reservoirs, leading to sedimentation. This process adversely affects drinking water sources, irrigation systems, and hydroelectric production. Sediments also carry pollutants such as pesticides and heavy metals, posing significant risks to ecosystems and human health (Toy et al., 2002).

Examples from Türkiye illustrate the seriousness of this issue. Many reservoirs across the country have experienced reduced storage capacity due to sedimentation over the past 50 years. This represents not only an economic concern but also a strategic environmental threat.

Loss of Biodiversity

Soil erosion affects not only soil organisms but also plant and animal species dependent on the soil, leading to a decline in terrestrial biodiversity. Microorganisms, fungi, earthworms, and various invertebrates are essential components that maintain soil structure and nutrient cycling. Erosion destroys the habitats of these organisms, thereby degrading soil health (García-Orenes et al., 2010).

Moreover, the disappearance of vegetation cover leads to the loss of feeding and breeding habitats for birds, insects, and mammals. The decline of pollinators and microfauna (microscopic soil fauna less than 100 μm in width, including nematodes and protozoa, which live in water-filled pores and feed on other soil microbes) can trigger cascading effects that ultimately collapse ecosystem services (Pimentel et al., 1995).

Climatic Feedback and Carbon Loss

Soil is one of the largest carbon reservoirs on Earth. Soils rich in organic matter play a critical role in carbon balance by absorbing atmospheric carbon and supporting microbial cycles. However, erosion converts this stored carbon into atmospheric form, increasing greenhouse gas emissions (Lal, 2004). According to the IPCC (2019), land degradation accounts for approximately 23% of global greenhouse gas emissions.

When erosion removes the organic-rich topsoil, the soil's carbon storage capacity declines, turning it from a carbon sink into a carbon source. This creates a positive feedback mechanism that amplifies the intensity of climate change (Lal, 2015).

Impact Area	Consequences of Erosion	Comparison / Regional Differences
Soil Fertility	Up to 50% yield loss (Pimentel & Burgess, 2013)	Europe: 1–2 t/ha/year; Africa: 30 t/ha/year (Panagos et al., 2015)
Water Quality	Up to 30% reservoir capacity loss due to sedimentation	Sediment control structures more effective in developed countries
Biodiversity	Loss of microorganisms and soil-dependent species	Faster decline in areas dominated by monoculture farming
Carbon Storage	Increased atmospheric CO_2, aggravating climate change	Losses are smaller in soils rich in organic matter

Erosion represents not only a surface-level soil loss problem but also a systemic risk to ecosystems. There is a strong interconnection between soil, water, biodiversity, and climate. Therefore, combating erosion is essential not only for agricultural productivity but also for maintaining the planet's ecological stability. Soil conservation measures, water management, agro-ecological practices, and reforestation—implemented as nature-based solutions—should be prioritized within this framework.

Socioeconomic Impacts of Erosion

Erosion is not only an environmental threat but also a fundamental cause of a chain of socioeconomic problems ranging from rural poverty to food crises, and from forced migration to urbanization pressure. In regions where livelihoods depend heavily on agriculture, erosion deepens social vulnerabilities by reducing productive lands and depleting natural resources (Lal, 2001; Pimentel et al., 1995). This process directly threatens the achievement of sustainable development goals.

Threat to Food Security

Erosion causes the loss of the most fertile layer of agricultural soils — the topsoil — thereby significantly reducing agricultural productivity. According to FAO (2020), a 1% decline in soil fertility can translate into the loss of millions of tons of grain globally. This loss can have particularly devastating consequences in regions where food security is already fragile, such as Africa, South Asia, and the Middle East.

For instance, in Sub-Saharan Africa, soil degradation has led to significant declines in productivity across 65% of agricultural lands (FAO, 2021). This not only triggers regional food crises but also contributes to price fluctuations, affecting global food markets (Montgomery, 2007). In Türkiye, erosion-induced soil loss has similarly resulted in measurable declines in agricultural production (TEMA, 2022).

The Poverty Cycle and Rural Economies

For rural communities, land is not merely a production asset but also the foundation of social and economic life. The reduction of arable land due to erosion threatens the livelihoods of millions of people living in rural areas. This results not only in income loss but also perpetuates intergenerational poverty, creating a self-reinforcing cycle (World Bank, 2019).

According to UNCCD (2022), in areas affected by land degradation, rural poverty rates are 35% higher compared to regions with healthier soils. In Türkiye, rural poverty rates are also above average in erosion-prone regions such as Eastern and Southeastern Anatolia. In this sense, erosion becomes a form of environmental injustice that directly contributes to the deepening of social inequalities.

Migration and Urbanization Pressure

In areas where agricultural production becomes unsustainable, people are forced to abandon rural lands after losing their means of livelihood. This process particularly drives younger populations toward cities, leading to rural depopulation and increasing challenges in urban areas such as unemployment, housing shortages, infrastructure pressure, and social integration problems (IOM, 2021).

Environmentally induced migration due to erosion, when combined with the impacts of climate change, may become one of the main triggers of mass population movements in the future. The IPCC (2019) projects that by 2050, nearly 200 million people worldwide may be displaced due to climatic and environmental factors. In Türkiye, increasing soil losses in Central, Eastern, and Southeastern Anatolia have accelerated rural-to-urban migration, aggravating unplanned urbanization problems in major cities such as İstanbul, Ankara, and İzmir.

Impact Area	Consequence of Erosion	Regional Variation / Comparison
Food Security	Yield loss, supply shortages, price increases	In Africa, productivity loss is 65%, while in Europe it remains around 20% due to sustainable practices (FAO, 2021).
Poverty	Loss of agricultural income, unemployment, intergenerational poverty transmission	Poverty is 35% higher in erosion-affected regions (UNCCD, 2022).
Migration & Urbanization	Rural depopulation, urban infrastructure stress, unemployment	Rural-to-urban migration is largely driven by environmental and livelihood factors.

The socioeconomic impacts of erosion clearly demonstrate that environmental degradation extends beyond nature and resources — it also threatens the structural stability of societies. Food production, livelihoods, and population dynamics are directly tied to soil health. Therefore, combating erosion is not only an environmental obligation but also closely linked to fundamental human concerns such as social justice, rural development, and the right to food. For policymakers and decision-makers, adopting this holistic perspective is of critical importance.

Erosion in Türkiye: Situation Analysis and Statistical Overview

Due to its geographical location, climatic characteristics, topography, and anthropogenic pressures, Türkiye is among the most erosion-prone regions in the world. Approximately 86% of the country's land area is under erosion risk — a rate nearly four times higher than the European average (FAO, 2021).

Regional Distributions

- Eastern Anatolia and Central Anatolia are the regions most affected by water erosion, primarily due to steep slopes, sparse vegetation, and intensive livestock grazing.
- The Black Sea Region experiences frequent landslides due to high rainfall intensity.

- Southeastern Anatolia is highly susceptible to wind erosion due to its semi-arid nature and irregular agricultural practices.

Erosion in Agricultural Lands

Moderate to severe erosion affects approximately 59% of Türkiye's agricultural lands (TÜİK, 2022). Traditional practices such as contour plowing, monoculture farming, and fallowing significantly accelerate erosion.

Region	Agricultural Land Affected by Erosion (%)
Central Anatolia	68%
Eastern Anatolia	72%
Southeastern Anatolia	55%
Black Sea	43%
Marmara	36%
Aegean	41%

Forests and Rangelands

A significant portion of Türkiye's forests is also affected by erosion. Additionally, 64% of rangelands exhibit evidence of soil displacement. Every year, Türkiye loses a considerable amount of fertile soil to wind and water erosion.

Local Knowledge and Traditional Practices

In many regions of Anatolia, traditional methods such as terracing, strip cropping, and perennial cultivation were once widely practiced as reflections of a deep-rooted soil conservation ethic. However, due to a lack of intergenerational transmission, much of this knowledge has been lost over time. The progressive disappearance of soil means not only ignoring the rights of present generations but also violating those of future ones. In this sense, erosion is also a matter of ecological justice.

Nature-Based Approaches in Combating Erosion

Efforts to combat erosion should focus not on transforming nature, but on living in harmony with it. Therefore, Nature-Based Solutions (NbS) have gained increasing prominence. In this context, vegetation restoration, reforestation, and afforestation are of key importance. Strengthening vegetation cover reduces surface runoff and binds soil particles. FAO (2021) reports that such methods can improve soil protection efficiency by 40–70%.

Additionally, terracing and water retention structures contribute significantly to erosion control. On sloped terrains, terracing enables the controlled movement of water, while stone walls and small ponds help slow down surface runoff. Participatory planning and the involvement of local communities are of great importance. Thus, erosion control is not only a technical issue but also a socio-cultural one. Integrating local knowledge and community needs into decision-making processes enhances both sustainability and effectiveness.

Sustainable Land Management (SLM) refers to the balanced and integrated use of soil, water, and vegetation resources. According to UNCCD (2022), regions implementing SLM practices have experienced a 25% increase in productivity and a 30% reduction in land degradation.

Global Perspective: Erosion, Climate Change, and Losses

Soil erosion is not merely a local problem but a global-scale crisis. Each year, approximately 24 billion tons of fertile soil are lost worldwide due to erosion (UNCCD, 2022). This loss directly threatens food security, ecosystem services, and the climate balance.

There is a linear and mutually reinforcing interaction between erosion and climate change. The transport of soil carbon increases the concentration of greenhouse gases in the atmosphere, while erosion simultaneously reduces the soil's water retention capacity, thereby accelerating aridification (Lal, 2004). This vicious cycle contributes to the rapid advancement of desertification.

International Policies and Goals

- United Nations Sustainable Development Goals (SDGs): Within the framework of SDG 15.3, which promotes Land Degradation Neutrality (LDN), the fight against erosion, ecosystem restoration, and nature-based solutions are prioritized (UN, 2015).
- UNCCD: Addresses soil erosion as part of a broader context that includes desertification, drought, and land degradation.
- FAO and IPBES: Jointly focus on land degradation linked to agricultural production and its implications for food security.

Economic Assessment of Erosion

Global Economic Losses

Soil erosion imposes a heavy economic burden on the global economy. According to reports by FAO (2021) and IPBES (2018):

- The impact on global agricultural productivity leads to an annual economic loss of 70–100 billion USD (Pimentel & Burgess, 2013).
- 33% of the world's agricultural land is moderately to severely degraded (FAO, 2021).
- Up to a 10% increase in global food prices may be attributed to erosion-induced production declines (IPBES, 2018).

"Erosion sweeps away not only soil, but also economies and lives." IPBES, 2018

Agricultural Production Losses

- Productivity losses range between 15–30%.
- Fertilizer and irrigation efficiency in eroded soils may decrease by up to 30%.
- Global grain production experiences annual losses of millions of tons due to soil erosion.

Sedimentation in Dams and Reservoirs

- The storage capacity of reservoirs decreases by 0.5% annually, meaning that 20% of a 40-year dam's lifespan is lost solely due to sedimentation.
- In 50 major dams, a total of 1.2 billion m³ of water storage capacity has been lost due to erosion-related sediment deposits.

Rehabilitation and Intervention Costs

- In Türkiye, the annual budget allocated to erosion control is approximately 2.5 billion TL (based on 2023 prices).
- However, due to the lack of long-term protective measures, the sustainability of these investments remains limited.
- According to the General Directorate of Forestry (OGM, 2022), the average cost of terracing per hectare is around 10,000 TL.

Erosion rates in Türkiye are significantly higher than the global average. In European countries such as Germany and France, soil conservation policies are systematic, integrated, and effectively enforced, while in Türkiye, despite existing legal frameworks, implementation remains fragmented. Therefore, a comparative evaluation provides valuable insight.

Comparative Indicators: Türkiye vs. Global Average

Indicators	**Türkiye**	**Global Average**
Area vulnerable to soil erosion	86%	33%
Annual soil loss	154–500 million tons	24 billion tons (global total)
Economic loss from erosion	~10 billion USD (estimated)	70–100 billion USD
Agricultural productivity loss	15–30%	10–20%
Reservoir capacity loss	0.5% per year	0.3–0.4% per year

Indirect Economic Impacts

- Food Price Increases: Reduced agricultural supply drives up consumer prices.
- Unemployment and Migration: Shrinking arable land reduces rural employment opportunities and accelerates rural-to-urban migration.
- Climate-Induced Disasters: Erosion amplifies the impacts of floods, droughts, and desertification, thereby increasing disaster-related costs.

Ten-Step Proposal for Sustainable Soil Management

1- *Land Use Planning:* Determine appropriate land uses based on slope, soil type, and climatic conditions.
2- *Terracing and Strip Cropping:* Implement especially in sloped terrains.
3- *Afforestation and Vegetation Programs:* Strengthen incentives for reforestation and green cover.
4- *Enhance Soil Organic Matter:* Enrich humus content to improve soil fertility.
5- *Adopt Sustainable Farming Practices:* Crop rotation, reduced tillage, and cover crops should be applied effectively.
6- *Controlled Pasture Management:* Prevent overgrazing and rehabilitate degraded grasslands.
7- *Reintegration of Traditional and Indigenous Knowledge:* Reassess local and ancestral land management techniques.
8- *Public Participation and Awareness:* Strengthen community engagement in decision-making processes.
9- *Legal and Institutional Frameworks:* Strengthen legislation and establish monitoring systems.
10- *Nature-Based Solutions and Ecosystem Restoration:* Support approaches that align human activity with natural processes.

Chapter Conclusion

Soil is not only the foundation of nature but also the cornerstone of human civilization. Erosion, however, is a silent and destructive process that threatens this foundation. Therefore, effective solutions must consider the physical, ecological, socioeconomic, and cultural dimensions of erosion.

Philosophical and Islamic perspectives emphasize that soil is not merely a resource, but also a moral and ethical entity entrusted to humanity. Hence, combating erosion must be addressed not only as a technical challenge, but as a matter of moral obligation, ethical responsibility, and intergenerational justice.

Protecting soil is, therefore, a spiritual and ethical duty as much as it is an environmental one. Policymakers at both national and global levels must develop strategic, data-supported, and awareness-raising frameworks to ensure the sustainability of soil — the very essence of life itself.

Chapter References

Ahmed b. Hanbel. (2001). *el-Müsned* (al-Musnad). Dâru'l-Fikr. Retrieved from https://iksadyayinevi.com/wp-content/uploads/2022/04/SIYER-ACISINDAN-AHMED-b.-HANBELIN-MUSNEDI-MEKKE-DONEMI.pdf

Ardoğan, R. (2013). İslam'da Çevre Teolojisinin Pratiğe Yansıması: Çevre Ahlakı (The Reflection of Environmental Theology in Islam: Environmental Ethics). *Birey ve Toplum Sosyal Bilimler Dergisi*, 2(1), 115-144.

Çelik, H. E., Topatan, H. & Akyüz, D. E. (2024). İklim Değişikliği, Orman Yangınları, Seller ve Gözlemler (Climate Change, Forest Fires, Floods, and Observations). Retrieved from https://www.researchgate.net/publication/385247326_Iklim_Degisikligi_Orman_Yanginlari_Seller_ve_Gozlemler

Çilek, A. & Berberoğlu, S. (2015). Konumsal Bilgi Sistemi Yardımıyla Türkiye'nin Erozyon Modellemesi (Erosion Modeling of Türkiye Using Geographic Information System). *Ç.Ü Fen ve Mühendislik Bilimleri Dergisi*, 33(6), 57-64.

Çullu, M. A., Şeker, H., Gözükara, G., Günal H. & Bilgili, A. V. (2024). Rapid characterization of soil horizons for different soil series utilizing Vis-NIR spectral information. *Geoderma Regional*, 38(1), 1-12. DOI: 10.1016/j.geodrs.2024.e00853

DSİ. (2023). 2023 Yılı Resmi Su Kaynakları İstatistikleri (Official Water Resources Statistics 2023). Ankara: Devlet Su İşleri Genel Müdürlüğü. Retrieved from:https://www.dsi.gov.tr/

Dutal, H. (2022). Determination of the impact of forest fires on soil erosion risk by using the ICONA model: a case study from the Mediterranean region of Turkey. *Turkish Journal of Forest Science*, 6(2), 510-538.

Ege, İ. (2025). Kur'an-ı Kerim'deki Ayetler Işığında Erozyon Konusuna Coğrafi Bakış (A Geographical Perspective on the Issue of Erosion in Light of the Qur'an Verses). *The Journal of Turk-Islam World Social Studies*, 44(44), 491-511. DOI: 10.29228/TIDSAD.78690

Erpul, G., & Saygın, S. D. (2012). Ülkemizde toprak erozyonu sorunu üzerine: Ne yapmalı? (On the Soil Erosion Problem in Our Country: What Should Be Done?) *Toprak Bilimi ve Bitki Besleme Dergisi*, 1(1), 26-32.

FAO. (2020). *The state of the world's land and water resources for food and agriculture – Systems at breaking point.* Food and Agriculture Organization of the United Nations. Retrieved from https://openknowledge.fao.org/server/api/core/bitstreams/ecb51a59-ac4d-407a-80de-c7d6c3e15fcc/content

FAO. (2021). *Global Soil Partnership: Soil Erosion Assessment.* Retrieved from https://www.fao.org/global-soil-partnership

FAO. (2021). *The state of the world's land and water resources for food and agriculture – Systems at breaking point.* Food and Agriculture Organization of the United Nations. https://doi.org/10.4060/cb9910en

FAO. (2022). *Global assessment of soil erosion and soil degradation.* Food and Agriculture Organization of the United Nations. Retrieved from https://www.fao.org

Froude, M. J., & Petley, D. N. (2018). *Global fatal landslide occurrence 2004–2016. Natural Hazards and Earth System Sciences*, 18(8), 2161–2181. https://doi.org/10.5194/nhess-18-2161-2018

García-Orenes, F., Roldán, A., Mataix-Solera, J., Cerdà, A., Campoy, M., Arcenegui, V., & Caravaca, F. (2010). *Soil structural stability and erosion rates influenced by agricultural management practices in a semi-arid Mediterranean agro-ecosystem. Soil Use and Management*, 26(3), 305–312. https://doi.org/10.1111/j.1475-2743.2010.00271.x

Goudie, A. S. (2014). *Arid and semi-arid geomorphology.* Cambridge University Press, UK.

Göl, C. & Tunca, M. (2025). Farklı arazi kullanım türü/örtüsü altındaki toprakların erozyon duyarlılıklarının değerlendirilmesi (Karataşbağı Deresi Havzası) (Evaluation of the Erosion Susceptibility of Soils under Different Land Use/Cover Types: Karataşbağı Stream Basin). *Anadolu Orman Araştırmaları Dergisi*, 11(1), 225-233. DOI: 10.53516/ajfr.1635440

Hooke, R. L. (2000). *On the history of humans as geomorphic agents. Geology*, 28(9), 843–846.
https://doi.org/10.1130/0091-7613(2000)28<843:OTHOHA>2.0.CO;2

IOM. (2021). *World Migration Report 2022.* International Organization for Migration. Retrieved from https://publications.iom.int/books/world-migration-report-2022

IPBES. (2018). *Land Degradation and Restoration: Summary for Policymakers.* Retrieved from
https://files.ipbes.net/ipbeswebprodpublicfiles/2018_ldr_full_report_book_v4_pages.pdf

IPCC. (2019). *Climate Change and Land: An IPCC special report on climate change, desertification, land degradation, sustainable land management, food security, and greenhouse gas fluxes in terrestrial ecosystems.* Retrieved from https://www.ipcc.ch/srccl/

İmamoğlu, A. & Dengiz, O. (2020). Komşu iki mikro havzada erozyon duyarlılık değerlerinin arazi örtüsü ve arazi kullanımına bağlı değişimin belirlenmesi (Determination of changes in erosion sensitivity values depending on land cover and land use in two neighboring micro-basins). *Toprak Bilimi ve Bitki Besleme Dergisi*, 8(1), 53-60. DOI: 10.33409/tbbbd.757017

Kamali, M. H. (2010). *The Middle Path of Moderation in Islam: The Qur'anic Principle of Wasatiyyah.* Oxford University Press, UK.

Karagöktaş, D. & Yakupoğlu, T. (2014). Erozyon araştırma sahasına dönüştürülmesi planlanan bir alanda aşınabilirlik ve toprak özellikleri arasındaki ilişkiler (Relationships between erodibility and soil properties in an area planned to be converted into an erosion research site). *Toprak Bilimi ve Bitki Besleme Dergisi*, 2(1), 6-12.

Karaoğlu, M. (2014). Erozyon, Rüzgâr Erozyonu ve Iğdır-Aralık Örneği (Erosion, Wind Erosion, and the Iğdır-Aralık Example). *Türk Tarım ve Doğa Bilimleri Dergisi*, 1(2), 167–172.

Lal, R. (2001). *Soil degradation by erosion. Land Degradation & Development*, 12(6), 519–539. https://doi.org/10.1002/ldr.472

Lal, R. (2004). *Soil carbon sequestration to mitigate climate change. Geoderma*, 123(1–2), 1–22. https://doi.org/10.1016/j.geoderma.2004.01.032

Lal, R. (2015). *Restoring soil quality to mitigate soil degradation. Sustainability*, 7(5), 5875–5895. https://doi.org/10.3390/su7055875

Montgomery, D. R. (2007). *Dirt: The Erosion of Civilizations.* University of California Press, US. Retrieved from https://books.google.com.tr/books/about/Dirt.html?id=HSu8r15nnoC&redir_esc=y

Montgomery, D. R. (2007). *Soil erosion and agricultural sustainability. Proceedings of the National Academy of Sciences*, 104(33), 13268–13272. https://doi.org/10.1073/pnas.0611508104

Morgan, R. P. C. (2005). *Soil erosion and conservation* (3rd ed.). Blackwell Publishing. Retrieved from https://bsssjournals.onlinelibrary.wiley.com/doi/abs/10.1111/j.1365-2389.2005.0756f.x

Nabhan, G. P. (1997). *Cultures of Habitat: On Nature, Culture, and Story.* Counterpoint. Retrieved from https://archive.org/details/culturesofhabita0000nabh

Naess, A. (1973). *The shallow and the deep, long-range ecology movement. Inquiry*, 16(1–4), 95–100.

Nasr, S. H. (1996). *Man and Nature: The Spiritual Crisis of Modern Man.* ABC International Group. Retrieved from https://books.google.com.tr/books/about/Man_and_Nature.html?id=lvEJAQAAMAAJ&redir_esc=y

OGM. (2022). *Orman Genel Müdürlüğü Faaliyet Raporu* [General Directorate of Forestry Annual Report]. Ankara: Orman Genel Müdürlüğü. Retrieved from https://www.ogm.gov.tr/tr

Öztürk, T., et al. (2020). *Fatal landslides in Turkey (1929–2019). Türk Coğrafya Dergisi*, 75, 55–68. Retrieved from https://dergipark.org.tr/tr/pub/tcd/issue/54666/731596

Panagos, P., Borrelli, P., Poesen, J., Ballabio, C., Lugato, E., Meusburger, K., Montanarella, L., & Alewell, C. (2015). *The new assessment of soil loss by water erosion in Europe. Environmental Science & Policy*, 54, 438–447. https://doi.org/10.1016/j.envsci.2015.08.012

Petley, D. (2024). *Fatal landslides in 2024. Eos*. Retrieved from https://eos.org/thelandslideblog/fatal-landslides-in-2024

Pimentel, D., & Burgess, M. (2013). *Soil erosion threatens food production. Agriculture*, 3(3), 443–463. https://doi.org/10.3390/agriculture3030443

Pimentel, D., Harvey, C., Resosudarmo, P., Sinclair, K., Kurz, D., McNair, M., Crist, S., et al. (1995). *Environmental and economic costs of soil erosion and conservation benefits. Science*, 267(5201), 1117–1123.

Ramazanoğlu, E., Almarie, V., Suzer, M. H., Shan, J., Wei, Z., Çullu, M. A., Bol, R. & Şenbayram, M. (2024). *Interactive effects of soil salinity and nitrogen fertilizer types on nitrous oxide and ammonia fluxes. Geoderma Regional*, 38, 1–10. https://doi.org/10.1016/j.geodrs.2024.e00831

Selby, M. J. (1993). *Hillslope materials and processes*. Oxford University Press, UK.

TEMA. (2022). *Faaliyet Raporu* [Activity Report]. Türkiye Erozyonla Mücadele Vakfı. Retrieved from https://www.tema.org.tr

Toy, T. J., Foster, G. R., & Renard, K. G. (2002). *Soil erosion: Processes, prediction, measurement, and control*. John Wiley & Sons. Retrieved from https://www.wiley.com/enus/Soil+Erosion%3A+Processes%2C+Prediction%2C+Measurement%2C+and+Control-p-9780471383697

TÜİK. (2021). *Göç İstatistikleri ve Kırsal Kalkınma Verileri* [Migration Statistics and Rural Development Data]. Retrieved from https://www.tuik.gov.tr/

TÜİK. (2022). *Tarım Çevre Göstergeleri* [Agricultural Environmental Indicators]. Türkiye İstatistik Kurumu. Retrieved from https://www.tuik.gov.tr/

TÜİK. (2022). *Tarımsal Yapı ve Üretim Verileri* [Agricultural Structure and Production Data]. Retrieved from https://www.tuik.gov.tr/

UNCCD. (2022). *Global Land Outlook 2: Land Restoration for Recovery and Resilience.* United Nations Convention to Combat Desertification. Retrieved from https://www.unccd.int/resources/global-land-outlook/glo2

United Nations. (2015). *Transforming our world: The 2030 Agenda for Sustainable Development.* Retrieved from https://sdgs.un.org/2030agenda

Uzun, S. M. (2025). *Antropojenik baskı yoğunluğu ile erozyon duyarlılığı ilişkisinin alt havza kapsamında analizi: Uluabat Gölü Havzası örneği* [Analysis of the relationship between anthropogenic pressure intensity and erosion susceptibility within a sub-basin: The case of Lake Uluabat Basin]. *Journal of Anatolian Geography*, 2(1), 48–62.

Wikipedia. (2024a). *2024 Wayanad landslides.* Retrieved from https://en.wikipedia.org/wiki/2024_Wayanad_landslides

Wikipedia. (2024b). *2024 Enga landslide.* Retrieved from https://en.wikipedia.org/wiki/2024_Enga_landslide

World Bank. (2019). *Ending Poverty and Boosting Shared Prosperity.* Retrieved from https://www.worldbank.org/en/topic/poverty

6th CHAPTER

DESERTIFICATION AS A GLOBAL THREAT

"In a world where consciences have turned into deserts, it is impossible to prevent the desertification of the soil."

H.E. Recep Tayyip ERDOĞAN, President of the Republic of Türkiye

Chapter Summary

This chapter examines the issue of desertification on both global and regional scales, analyzing its causes, consequences, geographic distribution, and strategies for combating it through a multidimensional framework. Factors such as drought, climate change, deforestation, improper land use, and excessive agricultural activities that accelerate desertification are explored in detail, with examples drawn from both global and Turkish contexts. Nature-based solutions, sustainable land management, international conventions, and policy recommendations are evaluated in terms of maintaining ecosystem services. Combating desertification requires not only technical interventions but also institutional reforms, policy integration, participatory planning, and financial mechanisms.

Keywords: Desertification, Land Degradation and Neutrality, Anti-Desertification Policies, International Initiatives

The Silent Exhaustion of Soil: A Multidimensional Analysis of Desertification

Throughout human history, the relationship established with nature has defined both the source and the limits of life. Yet each disruption of this relationship has inflicted wounds not only on the environment, nature, and natural resources but also on humanity itself. One of the most visible and perilous manifestations of this rupture is desertification. The loss of soil productivity should not be viewed merely as an ecological degradation process. Desertification is a multidimensional and profound crisis that deepens social injustices, increases poverty, triggers migration, and threatens ecosystem integrity.

Combating desertification is not solely a matter of soil; it is also about redefining humanity's relationship with nature and natural resources. Therefore, this issue is more than a technical analysis—it is a call to revisit the bond humanity shares with the Earth. Desertification is not just an environmental problem; it is the silent cry of soil, water, and life itself. It symbolizes not only the degradation of land but also the unraveling of the bonds between humans and nature. Beyond scientific realities, the current situation emphasizes that humanity must listen to nature and its resources—otherwise, the silenced voice of nature may one day silence humanity itself. Thus, combating desertification is not merely an act of intervention but a moral call rooted in responsibility and conscience.

Definition and Conceptual Framework of Desertification

Desertification is the process by which soils lose their productivity, typically occurring in arid, semi-arid, and dry sub-humid regions due to a combination of human activities and natural factors. According to the United Nations Convention to Combat Desertification (UNCCD), desertification is defined as "land degradation in arid, semi-arid, and dry sub-humid areas resulting from various factors, including climatic variations and human activities" (UNCCD, 1994).

Although the terms drought and desertification are often used interchangeably in everyday discourse, they refer to distinct phenomena. Drought is a temporary event characterized by below-average precipitation over a given period, whereas desertification denotes long-term and irreversible land degradation (Wilhite & Glantz, 1985). Desertification emerges when recurring droughts combine with natural or anthropogenic pressures, leading to lasting ecological damage. It is not solely an environmental problem but a complex process with economic, social, cultural, and political dimensions.

According to UNCCD (2022), 41% of the Earth's land surface consists of arid, semi-arid, and dry sub-humid regions, home to approximately 3.2 billion people. Since these communities largely depend on environmental and natural resource-based livelihoods, desertification poses both ecological and socio-economic threats. It is not merely the loss of soil productivity but a chain process leading to biodiversity loss, water scarcity, food insecurity, and forced migration (Safriel, 2007).

Global and Regional Distribution of Desertification

Approximately 35% of the global population lives in arid, semi-arid, and dry sub-humid regions (UNCCD, 2022). These areas represent the most vulnerable zones to desertification risk, with nearly 70% of them facing land degradation threats. Around 12 million hectares of land are lost annually due to desertification worldwide—equivalent to the disappearance of about 20,000 km² of arable land each day (UNCCD, 2022).

Globally, Africa is the continent most affected by desertification. The Sahel region, located south of the Sahara, is under intense pressure due to both climate change and overexploitation of land. The Sahel represents the epicenter of ecological and socio-economic crises related to desertification. According to FAO (2020), approximately 100 million people in this region are directly affected.

In Asia, China, the Central Asian Republics, and India; in the Middle East, Iran, Iraq, and Syria; and in the Arabian Peninsula, countries such as Saudi Arabia, Kuwait, the United Arab Emirates, and Qatar experience severe desertification. In the Americas, Mexico, Peru, and the southwestern United States face similar challenges. In China, the Gobi Desert expands by about 3,600 km² annually; in Central Asia, degradation around the Aral Sea has caused social and economic crises; and in India, more than 30% of the country's land is threatened by degradation (UNEP, 2017).

In Europe, Mediterranean countries such as Spain, Türkiye, and Greece are also at risk. Southern Spain and parts of Greece are particularly vulnerable due to declining irrigated farmland, deforestation, and ineffective land protection policies. Türkiye, located within the Mediterranean climate belt and increasingly affected by drought trends, is among the countries most exposed to desertification risk. Approximately 59% of Türkiye's land area is considered at risk (TÜİK, 2023).

According to the UNCCD Global Drought Hotspots Report (July 2025), desertification has evolved into a broad and severe global crisis. The report highlights critical drought hotspots across Africa (Somalia, Ethiopia, Zimbabwe,

Zambia, Malawi, Botswana, Namibia), the Mediterranean (Spain, Morocco, Türkiye), Latin America (Panama, the Amazon Basin), and Southeast Asia—emphasizing the urgency of desertification risks worldwide.

The threat of desertification affects not only rural but also urban and peri-urban areas due to pressures from urbanization, population growth, and overconsumption of natural resources. Excessive water use, urban sprawl, and the loss of green spaces necessitate an integrated rural-urban approach to combating desertification. Addressing this challenge therefore requires the harmonization of both rural and urban development policies.

Causes of Desertification: Natural Factors and Human Activities

The causes of desertification can generally be classified into two main categories: natural factors and anthropogenic (human-induced) factors. These factors often act together, influencing and reinforcing each other, thereby accelerating the process of desertification.

Among natural factors, climate change is the most prominent. Global warming increases evaporation rates, alters precipitation regimes, and leads to longer and more intense drought periods. According to IPCC (2021), the global average temperature has risen by approximately 1.1°C since 1880. This increase deepens water scarcity, particularly in arid and semi-arid regions, and accelerates soil desiccation by causing significant moisture loss. Moreover, sand and dust storms, which have become more frequent in recent years, further exacerbate the socio-economic impacts of desertification (Aydoğdu & Aydınoğlu, 2025).

According to the UNCCD (2025) report, the Mediterranean Basin, including Türkiye, is considered one of the most vulnerable regions to the impacts of climate change. Due to rising temperatures and declining rainfall, the Mediterranean has become one of the most affected areas where the consequences of global climate change are most visible. The Mediterranean climate, naturally characterized by dry summers, has experienced a significant increase in both the frequency and intensity of droughts since the 1950s. The report projects that average surface temperatures in the Mediterranean Basin will rise by 2–3°C by 2050 and 3–5°C by 2100. These temperature increases are expected to have direct effects on regional hydrological balances. Each 2°C increase in temperature is estimated to cause approximately a 15% reduction in water resources, posing substantial risks to agricultural production, ecosystem sustainability, and drinking water supply.

This fragile structure of the Mediterranean places the region at high risk not only ecologically but also socio-economically. Therefore, the development of integrated, climate-resilient policies tailored to the specific conditions of the Mediterranean region regarding desertification, land degradation, and water resource management is inevitable.

Physical factors such as soil characteristics, slope, and geological structure can also increase the risk of desertification. However, these factors often lead to more destructive consequences when combined with human activities.

Human activities, on the other hand, are the strongest drivers of desertification. The main contributors include:

- *Excessive agricultural activities:* Monoculture practices, intensive use of fertilizers and pesticides, and improper irrigation techniques degrade soil structure, increase erosion, and lead to salinization. According to FAO (2020), 33% of the world's agricultural lands are already degraded.
- *Deforestation:* Tree cutting and forest fires reduce vegetation cover, lower water retention capacity, and accelerate erosion. Between 1990 and 2020, the world lost 178 million hectares of forest land due to various causes (FAO, 2020).
- *Overgrazing:* Excessive and uncontrolled grazing reduces vegetation and ground cover. In semi-arid regions, this leads to soil compaction, loss of organic matter, and irreversible damage to plant regeneration capacity.
- *Overuse and mismanagement of water resources:* Excessive groundwater extraction and inefficient irrigation systems—particularly in closed basins—intensify salinization and land degradation. In Türkiye, for instance, the Konya Closed Basin is a striking example of this issue.
- *Improper land use and urbanization pressure:* Unplanned construction, conversion of agricultural land for settlement, and mining activities result in permanent soil degradation.

Most of these causes are interconnected, mutually reinforcing the desertification cycle. Therefore, combating desertification should not only rely on climatic perspectives but also incorporate socio-economic, cultural, and governance-based approaches.

Socio-Economic Consequences: Poverty, Migration, and Food Insecurity

Desertification is not only an environmental crisis but also a significant socio-economic threat. Low-income rural communities are disproportionately affected by this process, as their livelihoods directly depend on natural resources. According to the United Nations (UNCCD, 2022), approximately 135 million people experience loss of soil productivity each year due to desertification, leading to deepening poverty. Since agriculture, livestock, and other nature-based production systems are directly impacted, income loss and unemployment have become widespread (Aydoğdu & Canlı, 2025a).

The impacts of desertification on food security are also dramatic. FAO (2021) estimates that by 2050, agricultural production worldwide may decline by 10–25% due to desertification, water scarcity, and soil degradation. This situation will not only affect production but also increase food prices, disrupt distribution systems, and limit access to safe and sufficient food.

Migration is one of the most visible social consequences of desertification. In 2022, more than 32 million people were displaced due to climate and land degradation-related causes (IDMC, 2023). These migrations, often from rural to urban areas, increase the pressure on urban infrastructure and social services, resulting in further socio-economic challenges.

In the case of Türkiye, rising droughts in the Southeastern Anatolia, Central Anatolia, and Eastern Mediterranean regions have led to increasing rural migration in recent years due to soil productivity losses and declining agricultural incomes. This process has caused abandonment of rural areas, weakening of agricultural sustainability, and disintegration of the social fabric.

Socio-Economic Consequences of Desertification

Impact Area	**Tangible Outcomes**
Agriculture & Food	Decline in productivity, reduced crop diversity, food insecurity
Poverty	Income loss, depletion of livelihoods
Employment	Reduction in agricultural labor force, rural unemployment
Migration	Rural-to-urban migration, pressure on infrastructure

Social Structure	Breakdown of family structures, weakening of social bonds
Women & Vulnerable Groups	Economic marginalization due to land loss
Health	Sand and dust storms, water contamination, malnutrition
Education	School dropouts due to migration

For all these reasons, combating desertification is not merely an environmental issue but also a matter of poverty alleviation, social justice, and migration management.

The Social Justice Dimension of Desertification

Desertification is not only an environmental crisis but also a profound issue of social justice, equality, access to fundamental rights, and social vulnerability. Land degradation threatens not only natural systems but also the lives of people who depend directly on these systems. However, this threat does not affect all communities equally. The poor, women, local communities, smallholder farmers, and migrants are disproportionately impacted by desertification.

Inequality in Land Access and Property Rights

- Globally, land ownership remains largely in the hands of men. Although women contribute the most labor to land cultivation in many rural regions, they often lack formal ownership rights (FAO, 2020).
- Consequently, when land becomes degraded, women are often unable to access legal mechanisms for compensation or recovery.
- Similarly, local and indigenous communities frequently lack formal land rights over their ancestral territories, leaving them vulnerable to the impacts of desertification (UNDRIP, 2007).

Displacement, Migration, and Climate Injustice

- Each year, millions of people are displaced due to desertification, drought, and land degradation. These migrations often result in temporary shelters, insecure employment, and social exclusion (IDMC, 2023).

- Migrant communities face discrimination in accessing social services and are often forced to live in informal, unsafe, and precarious urban conditions (UN-Habitat, 2020).
- Those least responsible for the climate crisis are often those most affected by its consequences — a clear case of climate injustice (IPCC, 2019).

Inequality in Access to Livelihood Resources

- Smallholder farmers and women producers are often disadvantaged compared to large-scale agricultural enterprises in terms of access to technical support, irrigation systems, and financial resources (IFAD, 2019).
- Many desertification mitigation policies are designed to meet the needs of stronger, more influential actors rather than vulnerable groups, thereby deepening structural inequalities.

Representation Gaps in Decision-Making Processes

- Vulnerable and marginalized communities are rarely given adequate voice in shaping desertification-related policies.
- Decisions made without the participation of local people, women, and indigenous knowledge often lead to unsustainable and unjust outcomes (IUCN, 2020).
- Social participation is not only a right but also a necessity for effectiveness and ownership.

Environmental Justice and Intergenerational Responsibility

- Desertification jeopardizes future generations' access to fertile soil, clean water, and a safe environment.
- Thus, it violates not only the rights of today's people but also the rights of future generations, making it a matter of intergenerational justice (UNDP, 2021).

Harmony with Nature, Justice with Humanity

Combating desertification is not merely about conserving nature and natural resources; it is also about protecting human dignity, the right to life, and equality. True solutions must integrate social justice with environmental sustainability. Without the active participation of vulnerable groups, respect for local knowledge, and rights-based policies, efforts to combat desertification cannot achieve full success.

Therefore, every proposed solution must answer the questions: "For whom?" and "With whose participation?"

Impacts of Desertification on Vulnerable Groups: Women, Migrants, and Smallholder Farmers

The impacts of desertification are not experienced equally across society. Socioeconomically vulnerable groups — including women, the elderly, children, migrants, and small-scale farmers — suffer disproportionately. These groups are more vulnerable due to their direct dependence on natural resources, limited access to capital and services, and lack of representation in decision-making processes.

Women: The Invisible Owners of the Land

- Although women produce nearly 45% of agricultural output in developing countries, only 15% of them have legal land ownership (FAO, 2020).
- In Türkiye, agriculture is the sector with the highest female employment rate. In the TRC2 Region (Şanlıurfa and Diyarbakır), 59.2% of all employed women work in agriculture. Women are the backbone of agricultural production, often serving as unpaid family laborers, daily wage earners, or seasonal workers, under harsh conditions (Aydoğdu, 2019).
- Land degradation decreases women's productivity, limits their access to natural resources such as water, and increases their workload (UN Women, 2021).

- During drought periods, women in many regions walk over six hours per day to access water — a burden that also causes many girls to drop out of school (UNCCD, 2017).
- Women's participation in anti-desertification projects has been shown to increase success rates by 20–30% (IUCN, 2020).

Migrants and Displaced Populations

- Land degradation and drought are major drivers of forced migration. In 2022, over 32 million people were displaced for environmental reasons, most due to the collapse of agriculture-based livelihoods (IDMC, 2023).
- Migrants often work in informal, low-wage, and unsafe jobs in urban peripheries, with limited access to social protection systems (UN-Habitat, 2020).
- Female migrants face a dual vulnerability — both as women and as migrants — which places them at even greater risk (UNDP, 2021).

Smallholder Farmers: The Silent Victims of the Agricultural Crisis

- Around the world, 500 million smallholder farmers cultivate less than 2 hectares of land each, yet they produce 70% of the world's food (FAO, 2021).
- Desertification forces these farmers to lose their land, withdraw from production, and lose their livelihoods.
- A study conducted in the GAP-Şanlıurfa region of Türkiye found that 56.7% of farmers were unhappy with being farmers under current conditions (Aydoğdu et al., 2021a).
- Compared to large-scale farmers, smallholders have less access to credit, irrigation systems, and technical assistance, making them less resilient to climate change (World Bank, 2021).
- In many countries, smallholder farmers represent less than 10% of those benefiting from agricultural support programs (IFAD, 2019).

The Reality of Desertification in Türkiye: Trends and Statistics

Türkiye, located within the Mediterranean climate zone, is among the countries highly vulnerable to desertification due to its climatic, topographic, and anthropogenic conditions (Canlı & Aydoğdu, 2025). Approximately 59% of the country's land area is under the threat of desertification and land degradation (TÜİK, 2023). Processes such as droughts, soil erosion, water scarcity, and deforestation—particularly concentrated in the Central Anatolia, Southeastern Anatolia, Mediterranean, and Aegean regions—have made the impacts of desertification increasingly visible.

According to the National Desertification Risk Map studies conducted by the General Directorate of Combating Desertification and Erosion (ÇEM), which serves as Türkiye's national focal point for the UNCCD:

- Approximately 35% of Türkiye's land is at moderate to high risk of desertification,
- Erosion affects 59% of agricultural lands,
- Each year, around 1.4 million hectares of land are subject to degradation (ÇEM, 2021).

Drought Trends and Decline in Precipitation:

According to the Turkish State Meteorological Service (MGM), during the period 1971–2020, the country's annual average temperature increased by about 1.5°C, while average precipitation decreased by approximately 10% (MGM, 2022). The Southeastern Anatolia Region is one of the areas most affected by this decline; in 2023, it received 45% less rainfall than the long-term average (TÜİK, 2023). A survey conducted in Şanlıurfa within the GAP Region revealed that 57.7% of participants were concerned about climate change, 68.3% considered drought to be "fate," and 9.5% believed it was partially fate-related (Aydoğdu et al., 2021b). Therefore, scientifically grounded public awareness must be strengthened.

Konya Closed Basin: The Desertifying Heart of Türkiye

The Konya Closed Basin, one of Türkiye's largest agricultural production areas, has experienced severe deterioration in both agricultural productivity and environmental balance due to unsustainable water use, uncontrolled groundwater extraction, and climate change. Groundwater levels are rapidly declining, leading to the formation of thousands of sinkholes of varying sizes (SYGM, 2023).

Decline in Pasture and Forest Areas

In the 1950s, Türkiye had approximately 46 million hectares of pastureland, but today this figure has fallen below 14 million hectares (TÜİK, 2023). Although forest area has increased quantitatively, qualitative degradation has occurred. During the major forest fires of 2021, approximately 140,000 hectares of forest burned, releasing an estimated 8–10 million tons of CO_2 (TEMA, 2022).

Policy and Implementation Gaps in Combating Desertification

While Türkiye's strategic documents and action plans for combating desertification provide a solid policy framework, challenges persist during the implementation phase. These include lack of local-scale coordination, insufficient resources, and weak monitoring and evaluation mechanisms. Therefore, desertification in Türkiye should be considered not only an environmental issue but also an economic, social, and governance-related vulnerability.

International Approaches to Combating Desertification

Desertification is not only a national concern but also a critical global issue, remaining high on the international agenda. In this context, international strategies, conventions, cooperation models, and financial mechanisms developed to combat desertification are of great significance.

United Nations Convention to Combat Desertification (UNCCD):

Adopted in 1994 and entering into force in 1996, the UNCCD is the most comprehensive global convention addressing desertification, currently ratified by 197 countries. The Convention aims to combat land degradation in arid, semi-arid, and dry sub-humid areas while promoting sustainable land management (SLM) practices. Since 2015, the UNCCD has emphasized the concept of Land Degradation Neutrality (LDN), which seeks to ensure no net loss in overall land quality through the restoration of degraded lands (UNCCD, 2017; Aydoğdu & Canlı, 2025b).

Sustainable Development Goals (SDGs):

Within the framework of the United Nations 2030 Agenda, Goal 15: Life on Land is directly related to desertification. It aims to combat desertification, halt and reverse land degradation (UN, 2015; Aydınoğlu & Aydoğdu, 2025). The achievement of SDG-15 also supports other goals such as No Poverty (SDG-1), Zero Hunger (SDG-2), and Climate Action (SDG-13).

Intergovernmental Panel on Climate Change (IPCC) and Desertification:

In its 2019 Special Report, the IPCC highlighted the interlinkages between climate change and land degradation, emphasizing that desertification is directly connected to global warming, carbon emissions, and food security (IPCC, 2019). The report underscored the critical role of land management in climate action and recommended the implementation of nature-based solutions.

African Initiative: The Great Green Wall (GGW):

The Great Green Wall, launched by the African Union, is a multinational afforestation initiative across the Sahel region, aiming to establish an 8,000-kilometer green belt to combat desertification. The project targets the restoration of 100 million hectares of degraded land and the sequestration of 250 million tons of CO_2 by 2030 (UNCCD, 2021).

European Union Policies:

The European Union implements various funds and projects to support land protection, biodiversity, and agricultural sustainability—both within member states and in neighboring regions, particularly the Mediterranean Basin. The European Green Deal aims to reverse biodiversity loss and restore degraded ecosystems by 2030 (European Commission, 2020).

Financial Mechanisms and Global Funds:

Several international financial support mechanisms have been developed to combat desertification, including the Global Environment Facility (GEF), the Green Climate Fund (GCF), and the Adaptation Fund. Within the UNCCD framework, Türkiye has defined national LDN targets and initiated pilot projects in 14 watersheds to achieve them (ÇEM, 2021).

Policy Coordination and Monitoring Systems:

Combating desertification requires not only project-based interventions but also robust monitoring and evaluation systems. The Performance Review and Assessment of Implementation System (PRAIS) under the UNCCD enables countries to measure and compare progress. The Committee on Science and Technology (CST) provides scientific input for the implementation of the Convention, promoting research on land degradation processes and effective mitigation strategies (UNCCD, 2017). Established in 2013, the Science-Policy Interface (SPI) facilitates the flow of knowledge between scientists and policymakers, producing guidance documents to support the achievement of LDN targets (Chasek et al., 2019).

Policy Gaps and Solution Instruments

Policy Gap	Proposed Solution / Instrument
Fragmented legislation and institutional conflicts	Enact an integrated "Combating Desertification Law" and establish inter-agency coordination mechanisms.
Weak data infrastructure	Develop GIS-based national monitoring systems and UNCCD PRAIS-compatible data portals.
Limited local participation	Promote micro-watershed planning, inclusion of cooperatives and women's organizations in decision-making.
Financial insufficiency	Enhance access to GEF, GCF, and Adaptation Fund resources; establish national environmental funds.
Lack of education and awareness	Strengthen agricultural extension, media campaigns, and university collaborations.
Limited localization of technical practices	Expand nature-based solutions and traditional knowledge-based practices.
Weak implementation and monitoring	Establish performance-based monitoring systems and results-oriented auditing mechanisms.
Lack of coordination in water and soil management	Implement integrated water–soil allocation plans and promote the "farming according to water availability" model.

The transboundary nature of desertification makes international cooperation indispensable in addressing this crisis. Effective action requires not only technical solutions, but also financial, institutional, and political solidarity at both national and global scales.

The Impacts of Desertification on the Global Economy

Desertification is not merely an ecological crisis; it is also a rupture that can profoundly shake the global economy. Land degradation leads to declines in agricultural productivity, increases in food prices, loss of livelihoods, and economic migration, thereby creating macroeconomic instability on a global scale.

Impacts on Agricultural Production and the Food Economy

- Approximately 12 million hectares of land are lost every year due to desertification (UNCCD, 2022). This translates into an annual loss of about 20 million tons of food production—enough to meet the minimum annual grain or grain-equivalent needs (0.3 tons per person per year) of roughly 66–67 million people. In other words, this amount could feed the entire population of a country larger than the United Kingdom, France, Italy, and Spain combined (1.4 times) or about 3.7 times the Netherlands. This underscores how desertification poses a critical threat to global food security and human development.
- According to FAO (2020), 33% of global agricultural lands are already degraded.
- By 2050, global agricultural production is expected to decline by 10–25% due to desertification and land degradation (FAO, 2021).
- This reduction may lead to food price increases of up to 30%, severely exacerbating food insecurity in vulnerable regions.

Impacts on Employment and Livelihoods

- Around 1.5 billion people worldwide derive their livelihoods directly from the land (UNCCD, 2022).
- Desertification causes job losses and income reductions, particularly among small farmers, rural producers, and pastoral (nomadic livestock) communities.
- In the Sahel region of Africa alone, land degradation results in an annual economic loss of approximately USD 5 billion (UNEP, 2017).

Macroeconomic Losses and Impacts on Growth

- According to the World Bank (2021), land degradation causes about USD 44 billion in direct annual economic losses globally.
- When indirect impacts—such as food crises, health problems, migration, infrastructure damage, and ecosystem losses—are included, the total rises to over USD 300 billion (ELD Initiative, 2015).

- Land degradation can reduce GDP growth by up to 1.5% per year in low- and middle-income countries (UNDP, 2020).

Forced Migration and Impacts on Urban Economies

- In 2022, more than 32 million people were displaced due to climate- and land degradation-related causes (IDMC, 2023).
- These migrations exert pressure on urban labor markets, create housing shortages, and cause significant cost increases in infrastructure and public services.
- In some countries, budget allocations for urban infrastructure have increased by 20% due to rural-to-urban migration (UN-Habitat, 2020).

Economic Value and Loss of Ecosystem Services

- The annual economic value of ecosystem services such as soil carbon sequestration, water filtration, and erosion prevention is estimated at USD 125 trillion (Costanza et al., 2014).
- Desertification reduces the functionality of these services, leading to a significant decline in the world's natural capital stocks.

The Economic Reflection of the Ecological Crisis

Desertification negatively affects the global economic system both directly and indirectly—through production, employment, migration, food, and ecosystem services. This process particularly undermines development efforts in low-income countries, increases inequalities, and fuels global instability. Therefore, combating desertification must be viewed not only as an environmental issue but also as an integral part of economic development and welfare policies.

Methods for Combating Desertification: Nature-Based Solutions and Sustainable Land Management

In combating desertification, nature-based approaches are becoming increasingly important alongside classical engineering solutions. These approaches aim to restore ecosystem services through the protection, restoration, and sustainable management of natural processes. Nature-based solutions provide environmental, social, and economic benefits, thus building multidimensional resilience (IUCN, 2020).

Sustainable Land Management (SLM) is recognized as a fundamental strategy for preventing desertification and rehabilitating degraded lands. This management approach seeks to enhance long-term soil productivity by maintaining its physical, chemical, and biological properties.

Key SLM practices include:

- Appropriate tillage and minimal soil disturbance,
- Terracing on sloped lands,
- Organic matter enrichment,
- Improved irrigation efficiency,
- Promotion of perennial vegetation and agroforestry systems (UNCCD, 2022).

According to FAO (2021), SLM practices can increase productivity by 20–60% while reducing land degradation rates by up to 30%. In Türkiye, the General Directorate of Combating Desertification and Erosion (ÇEM), in cooperation with relevant institutions and funded by the Global Environment Facility (GEF), implements the project Integrated Natural Resource Management in the Very Humid Eastern Black Sea Region, coordinated by the UNDP and the Nature Conservation Centre, to promote sustainable and eco-friendly land management (ÇEM, 2025a).

Ecosystem-Based Adaptation

Ecosystem-Based Adaptation (EbA) is one of the most effective tools for climate adaptation within nature-based solutions. EbA enhances ecosystem services by increasing water retention capacity, reducing soil loss, and supporting biodiversity. In semi-arid regions, replanting drought-resistant species, supporting soil microorganisms, and preserving natural vegetation cover yield significant results in combating desertification.

Green Infrastructure and Landscape Planning

In areas under strong urbanization pressure, nature-based solutions should be supported with green infrastructure systems. City parks, xeriscaping adapted to each city's climate, green roofs, rain gardens, water harvesting systems, and permeable surfaces reduce erosion and improve water cycles.

In Türkiye, the General Directorate of Combating Desertification and Erosion has developed Woody Plant Guides Reflecting Urban Identity and Adapting to Climate Change, tailored to each province's climate and ecological characteristics, to strengthen urban green spaces (ÇEM, 2025b). This pioneering work represents a model contribution to national solutions.

The European Green Deal aims to increase urban green spaces by 25% by 2030 (European Commission, 2020). Urban greenery is an essential component of sustainable city life and directly affects citizens' right to live in a healthy environment. However, due to various factors, urban tree planting is unevenly distributed, causing certain neighborhoods to become more privileged in terms of green infrastructure. The concept of "tree justice" emphasizes equal access to the ecological, social, and health benefits of trees regardless of income, social status, or location. Therefore, urban forestry is not only an environmental sustainability measure but also a policy that guarantees equitable access to basic human rights (Yılmaz Güneş & Aydoğdu, 2025).

Community-Based Participation and Local Knowledge

The success of nature-based solutions is directly linked to the participation and knowledge of local communities. In combating desertification, the contributions of local farmers, women, and holders of traditional knowledge are of strategic importance. Strengthening community-based structures—such as education programs, cooperative models, and financial support mechanisms—significantly increases project success rates.

The Turkish Example: The Micro-Watershed Management Approach

Since the early 2000s, Türkiye has adopted integrated micro-watershed management projects. These projects balance land use, rehabilitate pastures and forests, and train farmers in sustainable production techniques. Examples include the GAP Action Plan, the Konya Closed Basin Management Plan, and erosion control projects by the General Directorate of Forestry (OGM, 2021).

The ÇEM General Directorate develops integrated micro-watershed rehabilitation projects aimed at the sustainable management of forests, pastures, soil, and water resources. These projects support economic and social activities that improve local welfare, reduce pressures on natural resources, and restore degraded ecosystem balance. In addition, they aim to mitigate natural disasters, adapt to the impacts of global warming, achieve land degradation neutrality, protect biodiversity, and increase soil organic carbon (ÇEM, 2025c).

Ecological Restoration and Biotechnical Applications

To repair degraded ecosystems within desertification processes, ecological restoration techniques are applied. These include:

- Rehydration of dried wetlands,
- Replanting of native species,
- Creation of biodiversity-supporting land mosaics,
- Biotechnical erosion control methods such as live fences, woven brushwood, and bio-barriers (SER, 2019).

Through such holistic approaches, both the biophysical environment and local livelihoods can be sustainably maintained. Nature-based solutions play a key role in enhancing climate resilience, strengthening carbon sinks, and slowing the progression of desertification.

Policy and Strategic Roadmap Recommendations

Combating desertification should not rely solely on technical interventions but must also be supported by holistic and inclusive policy frameworks. Presented below are several strategic policy recommendations that address the causes of desertification comprehensively, incorporating both preventive and restorative measures. These recommendations are also applicable to countries and regions with similar conditions.

1-Strengthening National Land Management Plans:

Policies for combating desertification should be more closely integrated with existing regional and national plans. Land-use planning must holistically and effectively address soil conservation, water resource management, and climate change adaptation together. The National Land Use Strategy should be made implementable at the local government level and better harmonized with sectoral plans.

2-Integrated Water and Soil Management:

One of the primary causes of desertification—unsustainable use of water and soil resources—should be addressed through integrated management approaches. In closed and water-scarce basins, water allocation plans must be regularly updated, the use of pressurized irrigation systems in agriculture should be expanded, and a water-based agricultural model should be adopted. Soil conservation measures such as erosion control, organic matter enhancement, and slope-appropriate cultivation must be further supported.

3-Strengthening Legal and Institutional Frameworks:

Policies on combating desertification should be consolidated under a coordinated framework law, rather than fragmented legal regulations. The current segmented legislative structure in areas such as forestry, rangelands, agriculture, water, and urbanization should be integrated, and a National Legislative Reform on Combating Desertification should be introduced. Institutionally, stronger coordination mechanisms should be established between relevant ministries and local governments, and National Desertification Coordination Councils should be formed.

4-Enhancing Monitoring, Early Warning, and Data Infrastructure:

Monitoring desertification processes, establishing early warning systems, and developing decision-support tools are essential. Remote sensing, satellite data, and GIS-based monitoring networks should be expanded, and publicly accessible Land Degradation Data Portals should be developed. The effectiveness of national monitoring systems aligned with the UNCCD's PRAIS (Performance Review and Assessment of Implementation System) should be enhanced.

5-Empowering Local Communities and Enhancing Participation:

Combating desertification is a field-based struggle. Therefore, the active participation of local populations—especially farmers, women, and youth—in decision-making and implementation processes should be promoted. Participatory micro-watershed planning, village-based rehabilitation projects, and cooperative supports should be further expanded.

6-Activating Financial Mechanisms:

National and international funding sources should be more effectively mobilized for nature-based solutions, sustainable agriculture, and restoration projects. Access to global funds such as the GEF (Global Environment Facility), GCF (Green Climate Fund), and the Adaptation Fund should be increased. At the national level, special environmental funds should be established for Land Degradation Neutrality. Agricultural support policies should be revised to promote production models that prevent desertification.

7-Education, Extension, and Awareness Programs:

Educational programs on desertification should target both rural and urban populations. Awareness seminars, digital learning platforms, and mass media tools should be used more intensively and effectively for agricultural advisors, teachers, public officials, NGOs, and youth. Universities and research institutions should establish strong interdisciplinary research networks on desertification.

8-Integration with Climate Change Policies:

Desertification policies should be effectively integrated with Nationally Determined Contributions (NDCs), the Climate Law, the Climate Adaptation Strategy, and all climate-related frameworks developed under the Paris Agreement. In doing so, combating land degradation will also contribute positively to enhancing carbon sinks, preserving biodiversity, and strengthening climate resilience.

In conclusion, success in combating desertification requires not only technical but also structural, administrative, and socio-cultural transformation. The policy recommendations outlined above will not only strengthen the capacity to combat desertification but also secure rural development and ecological integrity.

Chapter Conclusion

Desertification is a slow-moving yet devastatingly destructive process. It is not merely an environmental degradation issue but a global-scale threat that triggers social and economic vulnerabilities. Through its cascading effects—land degradation, biodiversity loss, depletion of water resources, declining soil fertility, food insecurity, and climate-induced migration—desertification poses a multifaceted threat to life. Both its causes and consequences consist of interlinked, self-reinforcing dynamics.

Since desertification transcends borders and affects the entire world, it cannot be effectively addressed through local-scale interventions alone. Therefore, developing joint mechanisms for collective action—supported by multilateral agreements and global cooperation—has become a necessity. In this context, the strategies and initiatives developed by the international community to combat desertification are of vital importance. These efforts focus on preventing land degradation and ensuring the sustainable use of soils. However, such international initiatives still face challenges, particularly regarding financial support, active participation of local communities, and the effectiveness of implementation policies.

For countries like Türkiye, a significant portion of which lies within semi-arid climatic zones, desertification directly influences multiple policy areas—from rural development to climate change mitigation. Examples such as the Konya Closed Basin demonstrate that desertification results not only from natural conditions but also from human-induced errors. Therefore, proposed solutions should be supported not only by technical interventions but also by institutional reforms, policy integration, participatory planning, and financial instruments.

Ultimately, combating desertification is not merely a problem of today—it is a matter of future security, ecological integrity, and intergenerational justice. Building a life model in harmony with nature, without depleting the opportunities it provides, is a shared responsibility of humankind.

The lasting traces of desertification are not only etched into the soil but also into the memory of humanity. Each statistic presented in this chapter represents a life; every barren land symbolizes a surrender. Combating desertification is not merely an environmental strategy—it is a moral obligation to future generations. As humanity continues to push the limits of nature, nature, in turn, reminds us of its own. If we fail to hear this voice today, our own voices may fall silent tomorrow. Therefore, this chapter stands not only as a scientific statement but also as an inner warning.

Chapter References

Altunsu, M., Oğuz, I., & Koçyiğit, R. (2021). *Kurak ve Farklı Eğim Koşullarına Sahip Bir Arazi Toplulaştırma Sahasının Çölleşme Potansiyelinin Klasik İstatistiksel Yöntemlerle Araştırılması* [Investigation of the Desertification Potential of a Land Consolidation Area with Arid and Different Slope Conditions Using Classical Statistical Methods]. *Bartın Orman Fakültesi Dergisi*, 23(1), 285–293. https://doi.org/10.24011/barofd.781727

Atabey, E. (2024). *Çölleşme ve İklim Değişikliği* [Desertification and Climate Change]. Retrieved from https://www.researchgate.net/publication/377737049_COLLESME_VE_IKLIM_DEGISIKLIGI

Aydınoğlu, E., & Aydoğdu, M. H. (2025). *Arazi Bozulumunu Dengeleme Mücadelesi Süreci: Kumaş Aynı Modeller Farklı* [The Process of Combating Land Degradation Neutrality: Same Fabric, Different Models]. *ICHEAS 9th International Conference on Health, Engineering and Applied Sciences*, March 14–16, Casablanca, Morocco, 82–93. ISBN: 978-625-5962-51-5

Aydoğdu, M. H., & Aydınoğlu, E. (2025). *Kum ve Toz Fırtınalarının Kökeni: Doğal mı, İnsan Kaynaklı mı?* [The Origin of Sand and Dust Storms: Natural or Anthropogenic?]. *Latin America 10th International Conference on Scientific Researches*, May 1–4, Havana, Cuba, 451–457. ISBN: 978-625-5962-70-6

Aydoğdu, M. H., & Canlı, P. (2025a). *Çölleşmeyle Mücadelede Küresel Perspektif: Sorunlar, Kurumsal Yaklaşımlar ve Sürdürülebilir Çözüm Önerileri Üzerine Genel Bir Değerlendirme* [Global Perspective in Combating Desertification: A General Assessment on Problems, Institutional Approaches and Sustainable Solution Proposals]. *Latin America 10th International Conference on Scientific Researches*, May 1–4, Havana, Cuba, 443–450. ISBN: 978-625-5962-70-6

Aydoğdu, M. H., & Canlı, P. (2025b). *Arazi Tahribatı: Küresel Bir Tehdit Olarak Önemi, Etkileri ve Sürdürülebilir Çözüm Yaklaşımları* [Land Degradation: Its Importance, Impacts, and Sustainable Solution Approaches as a Global Threat]. *Latin America 10th International Conference on Scientific Researches*, May 1–4, Havana, Cuba, 435–442. ISBN: 978-625-5962-70-6

Aydoğdu, M. H. (2019). *Tarım sektöründe çalışan kadınların işe bakışları ve memnuniyetleri üzerine bir araştırma* [A Study on the Job Perceptions and Satisfaction of Women Working in the Agricultural Sector]. *Harran Tarım ve Gıda Bilimleri Dergisi*, 23(4), 380–390. https://doi.org/10.29050/harranziraat.555463

Aydoğdu, M. H., Cançelik, M., Sevinç, M. R., Çullu, M. A., Yenigün, K., Küçük, N., Karlı, B., Ökten, Ş., Beyazgül, U., Doğan, H. P., Sevinç, G., Şahin, Z., Mutlu, N., Kaya, C., Yenikale, A., & Yenikale, A. (2021a). *Are You Happy to Be a Farmer? Understanding Indicators Related to Agricultural Production and Influencing Factors: GAP-Şanlıurfa, Turkey. Sustainability*, 13(22), 12663. https://doi.org/10.3390/su132212663

Aydoğdu, M. H., Cançelik, M., Sevinç, M. R., Çullu, M. A., Yenigün, K., Küçük, N., Karlı, B., Ökten, Ş., Beyazgül, U., Doğan, H. P., Şahin, Z., Mutlu, N., Kaya, C., Yenikale, A., & Yenikale, A. (2021b). *Is Drought Caused by Fate? Analysis of Farmers' Perception and Its Influencing Factors in the Irrigation Areas of GAP-Şanlıurfa, Turkey. Water*, 13(18), 2519. https://doi.org/10.3390/w13182519

Canlı, P., & Aydoğdu, M. H. (2025). *Türkiye'nin Çölleşmeyle Mücadele Stratejisi* [Turkey's Strategy for Combating Desertification]. *ICHEAS 9th International Conference on Health, Engineering and Applied Sciences*, March 14–16, Casablanca, Morocco, 44–55. ISBN: 978-625-5962-51-5

Chasek, P., Safriel, U., Shikongo, S., & Helgeson, J. (2019). *Operationalizing Land Degradation Neutrality: The Next Phase. Sustainability*, 11(3), 679. https://doi.org/10.3390/su11030679

Costanza, R., de Groot, R., Sutton, P., van der Ploeg, S., Anderson, S. J., Kubiszewski, I., ... & Turner, R. K. (2014). *Changes in the global value of ecosystem services. Global Environmental Change*, 26, 152–158. https://doi.org/10.1016/j.gloenvcha.2014.04.002

ÇEM. (2021). *Türkiye Arazi Bozulumu Nötrlüğü Hedef Belirleme Raporu* [Turkey's Land Degradation Neutrality Target Setting Report]. Ankara: Çölleşme ve Erozyonla Mücadele Genel Müdürlüğü. Retrieved from https://cem.csb.gov.tr/

ÇEM. (2021). *Ulusal Çölleşme Risk Haritası* [National Desertification Risk Map]. Ankara: Çölleşme ve Erozyonla Mücadele Genel Müdürlüğü. Retrieved from https://cem.csb.gov.tr/

ÇEM. (2025a). *DOKSAY Projesi* [DOKSAY Project]. Ankara: Çölleşme ve Erozyonla Mücadele Genel Müdürlüğü. Retrieved from https://cem.csb.gov.tr/doksay-projesi-baslangic-calistayi-rize-de-gerceklestirildi-haber-291303

ÇEM. (2025b). *Bitki Rehberleri* [Plant Guides]. Ankara: Çölleşme ve Erozyonla Mücadele Genel Müdürlüğü. Retrieved from https://cem.csb.gov.tr/bitki-rehberleri-i-112910

ÇEM. (2025c). *Havza Planlama ve Arazi Islahı Projeleri* [Watershed Planning and Land Rehabilitation Projects]. Ankara: Çölleşme ve Erozyonla Mücadele Genel Müdürlüğü. Retrieved from https://cem.csb.gov.tr/havza-planlama-ve-arazi-islahi-i-113012

Dölarslan, M., & Gül, E. (2024). *Çölleşme ve Arazi Bozulumu: Kavramsal Olgular ve Modelleme* [Desertification and Land Degradation: Conceptual Facts and Modelling]. Livre de Lyon. ISBN: 978-2-38236-678-3

Eker, Ö. (2015). *Çölleşme ve Kuraklığın Göç ve Yoksulluk Üzerine Etkileri: Yeniyapan Köyü Örneği* [The Effects of Desertification and Drought on Migration and Poverty: The Case of Yeniyapan Village]. Retrieved from https://www.researchgate.net/publication/295819515_Collesme_ve_Kurakligin_Goc_ve_Yoksulluk_Uzerine_Etkileri_Yeniyapan_Koyu_Ornegi

ELD Initiative. (2015). *The Value of Land: Prosperous lands and positive rewards through sustainable land management.* Bonn: Economics of Land Degradation. Retrieved from https://www.eld-initiative.org

European Commission. (2020). *EU Biodiversity Strategy for 2030.* Brussels: European Commission. Retrieved from https://environment.ec.europa.eu/strategy/biodiversity-strategy-2030_en

FAO. (2020). *Gender and Land Rights Database.* Rome: Food and Agriculture Organization of the United Nations. Retrieved from https://openknowledge.fao.org/items/a85a029e-4955-4640-9d48-3ef16aed75de

FAO. (2020). *The State of the World's Forests 2020.* Rome: Food and Agriculture Organization of the United Nations. Retrieved from https://www.fao.org/state-of-forests/en/

FAO. (2021). *The State of Food and Agriculture 2021: Making agrifood systems more resilient to shocks and stresses.* Rome: Food and Agriculture Organization. Retrieved from https://openknowledge.fao.org/items/437c1215-556b-4161-9af6-68163f5a1f84

IDMC. (2023). *Global Report on Internal Displacement 2023.* Geneva: Internal Displacement Monitoring Centre. Retrieved from https://www.internal-displacement.org/global-report/grid2023/

IFAD. (2019). *Creating Opportunities for Rural Youth: 2019 Rural Development Report.* Rome: International Fund for Agricultural Development. Retrieved from https://www.ifad.org/en/w/publications/2019-rural-development-report

IPCC. (2019). *Climate Change and Land: An IPCC Special Report.* Geneva: Intergovernmental Panel on Climate Change. Retrieved from https://www.ipcc.ch/site/assets/uploads/2019/11/SRCCL-Full-Report-Compiled-191128.pdf

IPCC. (2021). *Intergovernmental Panel on Climate Change, Sixth Assessment Report.* Geneva: Intergovernmental Panel on Climate Change. Retrieved from https://www.ipcc.ch/assessment-report/ar6/

Işık, S. (2024). *Büyüyen Çöl ya da Hiçliğin Çölü(nde): Nietzsche'de Çöl İmgesinin Kullanımı Üzerine* [The Growing Desert or the Desert of Nothingness: On the Use of the Desert Image in Nietzsche]. *Journal of Literature and Humanities.* https://doi.org/10.55590/literatureandhumanities.1464237

IUCN. (2020). *Gender and Nature-based Solutions.* Gland: International Union for Conservation of Nature. Retrieved from https://portals.iucn.org/library/sites/library

IUCN. (2020). *Nature-based Solutions Global Standard.* International Union for Conservation of Nature. Retrieved from https://portals.iucn.org/library/sites/library/files/documents/2020-020-En.pdf

Kılınç, B. (2024). *Sosyolojik Bakışla Çölleşmenin Topluma Yansıyan Etkileri* [Sociological Perspective on the Social Impacts of Desertification]. *Sosyoloji Dergisi.* https://doi.org/10.59572/sosder.1386934

Metinoğlu, F., Doğan, Z. A., & Özdemir, I. (2014). *Türkiye'de Kuraklık ve Çölleşmenin Kırsal Göç'e Etkileri* [The Effects of Drought and Desertification on Rural Migration in Turkey]. Retrieved from https://www.researchgate.net/publication/282356553_Turkiye'de_Kuraklik_ve_Collesmenin_Kirsal_Goc'e_Etkileri

MGM. (2022). *Türkiye Yıllık İklim Değerlendirmeleri* [Turkey Annual Climate Assessments]. Ankara: Meteoroloji Genel Müdürlüğü. Retrieved from https://www.mgm.gov.tr/iklim/iklim-raporlari.aspx

Mutlu, N., Gunal, H., & Acir, N. (2013). *Çölleşme; Nedenleri, Belirlenmesi ve İzlenmesi* [Desertification: Its Causes, Determination and Monitoring]. Retrieved from https://www.researchgate.net/publication/281111543_Collesme_Nedenleri_Belirlenmesi_ve_Izlenmesi

OGM. (2023). *Baraj Havzaları Erozyon Kontrolü Eylem Planı (2024-2028).* Ankara: Orman Genel Müdürlüğü. [General Directorate of Forestry (2023). *Erosion Control Action Plan for Dam Basins (2024–2028).* Ankara: General Directorate of Forestry.] Retrieved from https://www.ogm.gov.tr/tr/e-kutuphane-sitesi/Yayinlar/Baraj%20Havzalar%C4%B1%20Erozyon%20Kontrol%C3%BC%20Eylem%20Plan%C4%B1%20(2024-2028).pdf

OGM. (2023). *Orman Yangınları Değerlendirme Raporu.* [General Directorate of Forestry (2023). *Forest Fires Assessment Report.*] Retrieved from https://www.ogm.gov.tr/muglaobm/duyurular-sitesi/Documents/mugla-obm-2023yiliyangindegerlendirmeraporu/2023+Yili+Orman+Yanginlari+Degerlendirme+Raporu.pdf

Özüpekçe, S. (2020). *Güneydoğu Anadolu Bölgesinde Kuraklık, Çölleşme ve Tarımsal Ürün Verimi Arasındaki İlişkiler. [The Relationships Between Drought, Desertification, and Agricultural Productivity in Southeastern Anatolia Region.] Turkish Studies - Social Sciences, 15*(8), 3713–3726. DOI: 10.47356/TurkishStudies.43859

Özyol, K. (2024). *Çölleşmenin Ekosisteme Etkileri ve Çölleşmeyi Tersine Çevirme Yolunda Sürdürülebilir Tarımın Önemi.* [*The Effects of Desertification on the Ecosystem and the Importance of Sustainable Agriculture in Reversing Desertification.*] *Anadolu Orman Araştırmaları Dergisi, 8*(1), 113–122. DOI: 10.53516/ajfr.1060466

Safriel, U. (2007). *The Assessment of Global Trends in Land Degradation.* In D. J. Greenland et al. (Eds.), *Soil Quality and Agricultural Sustainability.* CRC Press, US.

Sağlam, A. (2024). *Çevresel Sorunlar İnsan Sağlığına, Ekonomilere ve Dünyaya Zarar Vermeye Devam Ediyor. Kadim Akademi [Environmental Problems Continue to Harm Human Health, Economies, and the Planet.] SBD, 8*(2). DOI: 10.55805/kadimsbd.1553015

SER. (2019). *Standards for Ecological Restoration.* Society for Ecological Restoration. Retrieved from https://www.ser.org/page/serstandards/international-standards-for-the-practice-of-ecological-restoration.htm

SYGM. (2023). *Konya Havzası Kuraklık Yönetim Planının Güncellenmesi Projesi.* Ankara: Su Yönetimi Genel Müdürlüğü. [General Directorate of Water Management (2023). *Project for Updating the Konya Basin Drought Management Plan.* Ankara: GDWM.] Retrieved from https://webdosya.csb.gov.tr/db/scd/icerikler/n-ha--plan-raporu-20231019105747.pdf

TÜİK. (2023). *Türkiye İstatistik Kurumu Çevre İstatistikleri.* [Turkish Statistical Institute (2023). *Environmental Statistics of Turkey.*] Retrieved from https://www.tuik.gov.tr/

TÜİK. (2023). *Türkiye İstatistik Kurumu Çevre ve Tarım Verileri.* [Turkish Statistical Institute (2023). *Environmental and Agricultural Data of Turkey.*] Retrieved from https://www.tuik.gov.tr/

Türkeş, M. (2013). *İklim Verileri Kullanılarak Türkiye'nin Çölleşme Haritası Dokümanı Hazırlanması Raporu.* [*Report on the Preparation of Turkey's Desertification Map Using Climate Data.*] Retrieved from https://www.researchgate.net/publication/293334692_Iklim_Verileri_Kullanilarak_Turkiye'nin_Collesme_Haritasi_Dokumani_Hazirlanmasi_Raporu

Türkeş, M. (2015). *Kuraklık, Çölleşme ve Birleşmiş Milletler Çölleşme İle Savaşım Sözleşmesi'nin Ayrıntılı Bir Çözümlemesi.* [*A Detailed Analysis of the United Nations Convention to Combat Desertification and Its Relevance to Drought and Desertification.*] *Marmara Üniversitesi Avrupa Topluluğu Enstitüsü Avrupa Araştırmaları Dergisi, 20*(1), 7–55.

Türkeş, M. (2017). *Birleşmiş Milletler Çölleşme ile Savaşım Sözleşmesi ve Türkiye'nin Durumu-Uyumu.* [*The United Nations Convention to Combat Desertification and Turkey's Status and Compliance.*] Retrieved from https://www.researchgate.net/publication/328916807_Birlesmis_Milletler_Collesme_ile_Savasim_Sozlesmesi_ve_Turkiye'nin_Durumu-Uyumu

Türkeş, M. (2022). *17 Haziran Dünya Çölleşme ve Kuraklıkla Savaşım Günü: Kuraklıktan birlikte yükselmek.* [*June 17 World Day to Combat Desertification and Drought: Rising Together from Drought.*] Retrieved from https://www.researchgate.net/publication/361419058_17_Haziran_Dunya_Collesme_ve_Kuraklikla_Savasim_Gunu_Kurakliktan_birlikte_yukselmek

Türkeş, M. (2023). *Kuraklık Yönetimi ve Kuraklık Afet Riskinin Azaltılmasına İlişkin Temel Kavramlar ve Önlemler-Öneriler.* [*Drought Management and Basic Concepts, Measures, and Recommendations for Reducing Drought Disaster Risk.*] Retrieved from https://www.researchgate.net/publication/369641640_Kuraklik_Yonetimi_ve_Kuraklik_Afet_Riskinin_Azaltilmasina_Iliskin_Temel_Kavramlar_ve_Onlemler-Oneriler

UN Women. (2021). *The Gender Gap in Land Ownership.* New York: United Nations Entity for Gender Equality and the Empowerment of Women. Retrieved from https://www.unwomen.org/en/digital-library/publications/

UN. (2015). *Transforming our world: The 2030 Agenda for Sustainable Development.* United Nations General Assembly. Retrieved from https://www.un.org/en/

UNCCD. (1994). *United Nations Convention to Combat Desertification.* Retrieved from https://www.unccd.int/

UNCCD. (2017). *Gender and Desertification: Expanding Roles for Women to Restore Land and Soil.* Bonn: United Nations Convention to Combat Desertification. Retrieved from https://www.unccd.int/

UNCCD. (2017). *Scientific Conceptual Framework for Land Degradation Neutrality.* Bonn: United Nations Convention to Combat Desertification. https://www.unccd.int/

UNCCD. (2021). *Great Green Wall: Implementation Status and Way Ahead to 2030.* Bonn: UNCCD Secretariat. Retrieved from https://www.unccd.int/

UNCCD. (2022). *Global Land Outlook 2nd Edition.* Bonn: United Nations Convention to Combat Desertification. Retrieved from https://www.unccd.int/

UNCCD. (2025). *Global Drought Hotspots Report Catalogs Severe Suffering, Economic Damage.* Retrieved from https://www.unccd.int/news-stories/press-releases/global-drought-hotspots-report-catalogs-severe-suffering-economic

UNDP. (2020). *Human Development Report 2020: The Next Frontier—Human Development and the Anthropocene.* New York: United Nations Development Programme. Retrieved from https://www.undp.org/

UNDP. (2021). *Human Development Report 2021/22: Uncertain Times, Unsettled Lives.* New York: United Nations Development Programme. Retrieved from https://www.undp.org/

UNDP. (2021). *Nature-Based Solutions for Land Management.* New York: United Nations Development Programme. Retrieved from https://www.undp.org/

UNDRIP. (2007). *United Nations Declaration on the Rights of Indigenous Peoples.* New York: United Nations. Retrieved from https://www.un.org/development/desa/indigenouspeoples/wp-content/uploads/sites/19/2018/11/UNDRIP_E_web.pdf

UNEP. (2017). *Global Environmental Outlook 6 - Regional Assessments: Africa.* Nairobi: United Nations Environment Programme. Retrieved from https://www.unep.org/

UN-Habitat. (2020). *World Cities Report 2020: The Value of Sustainable Urbanization.* Nairobi: United Nations Human Settlements Programme. Retrieved from https://unhabitat.org/world-cities-report-2020-the-value-of-sustainable-urbanization

Wilhite, D. A., & Glantz, M. H. (1985). *Understanding the Drought Phenomenon: The Role of Definitions. Water International, 10*(3), 111–120.

World Bank. (2021). *The Changing Wealth of Nations 2021: Managing Assets for the Future.* Washington, DC: World Bank. https://doi.org/10.1596/978-1-4648-1590-4

Yılmaz Güneş, A., & Aydoğdu, M. H. (2025). *Şehirlerin Yeşil Dönüşümü: Çevre Hakkı Perspektifinden Ağaç Eşitliği.* [*The Green Transformation of Cities: Tree Equality from the Perspective of Environmental Rights.*] *ICHEAS 9th International Conference on Health, Engineering and Applied Sciences*, March 14–16, Casablanca–Fas, s. 34–43. ISBN: 978-625-5962-51-5.

7th CHAPTER

LAND DEGRADATION: A GLOBAL THREAT TO THE FRAGILITY OF NATURE AND THE "BALANCING" APPROACH

Chapter Summary

Soil… it is not merely the ground we walk upon, but the memory of the past, the carrier of the present, and the guarantor of the future. As "modern" humanity weakens its bond with nature, the one thing it has most profoundly forgotten is the voice of the soil. Yet, even in silence, the soil speaks—through its cracks, its stillness, and the exhaustion accumulating within its increasingly barren body. This chapter does not describe merely an environmental issue; it tells of a civilizational crisis, a moral rupture, and a struggle for life itself. Here, "land degradation" is not addressed only as a technical or environmental deterioration but is questioned within philosophical, legal, social, and cultural dimensions. It reminds us of the forgotten covenant between humankind and the land. The chapter stands as an ethical call to nature, a holistic proposal for policy, and a moral question posed to humanity: "Do we wish to continue our war against the soil, or shall we learn once more to live in peace with it?" Every concept, every datum, and every idea presented here is not only an academic contribution but also a call to life itself. And now… it is time to hear the soil that screams in silence — and to think anew.

***Keywords*:** Land Degradation, Sustainable Land Management, Ecological Ethics and Soil Rights, Land Degradation Neutrality (LDN), Rural Poverty, and Land Justice

The Necessity of Harmony with Nature

Throughout human history, the relationship established with nature and natural resources has not only formed the foundation of life itself but has also played a decisive role in shaping civilizations. At the heart of this relationship lies land —

not merely as an economic resource but as the bearer of ecological balance, cultural identity, and social sustainability.

However, as a consequence of rapid population growth, urbanization, industrialization, and intensified agricultural activity over the past century, the ancient balance between humans and nature has been profoundly disrupted. The soil is no longer seen solely as a productive surface; it has become a global element of threat and vulnerability.

Land degradation is now regarded not simply as an environmental issue but as a complex and multidimensional crisis. Degraded soils, desertified lands, and collapsing ecosystems send us not only an environmental message but also a moral and philosophical warning:

"If you are at war with nature, you cannot be the winner."

Therefore, the issue of land degradation addressed in this chapter goes beyond describing environmental deterioration; it critically examines our relationship with nature and natural resources, questions their inherent rights, and challenges the limits of anthropocentric (human-centered) thinking.

The concept of Land Degradation Neutrality (LDN) represents a response developed to confront this crisis — a model that integrates both technical and ethical dimensions. Framed within the United Nations Sustainable Development Goals (SDG 15), this approach compels us to view land not merely as an economic commodity but as a living space, a right, and a being in its own regard.

The soil is the memory of nature. Beneath the ground we walk upon lie countless layers of life that we have forgotten. To protect it, therefore, is to safeguard not only the future but also the past and the present.

What Is Land Degradation? A Conceptual and Philosophical Approach

Conceptual Definition and Scope

Land degradation is defined as the process in which the natural capacity of land resources — particularly soil, water, vegetation, and biodiversity — declines, leading to the deterioration of the ecosystem services they provide (UNCCD, 2015). In other words, land degradation refers to the biophysical, chemical, hydrological, and biological deterioration of soil properties — processes that are often difficult or even impossible to reverse (IPBES, 2018). Land degradation occurs not only on agricultural lands but also in forests, rangelands, wetlands, and even in urban areas. The major factors driving this process include inappropriate land use, intensive agricultural practices, deforestation, overgrazing, mining, and climate change (FAO & UNEP, 2021). In particular, land-use change and the loss of soil organic carbon are both causes and indicators of degradation.

The Land–Degradation–Technology Relationship

Historically, technological progress has often functioned as a tool that tests — and exceeds — the limits of the soil. Industrial agricultural techniques, the use of chemical fertilizers and pesticides, and monoculture systems (the long-term cultivation of a single crop species across large areas) may increase yields in the short term but, over time, disrupt the soil's natural structure. These practices negatively affect carbon cycles, water regimes, and soil microbiota — the microorganisms that sustain plant growth, provide disease resistance, and enhance soil fertility (Lal, 2019).

Therefore, land degradation is not only an environmental problem but also a problem of knowledge and morality. What we produce is important — but how we produce it is even more so. Production systems that deplete the capacity of nature and its resources ultimately amount to the theft of the future itself (Ostrom, 2009).

Philosophical Perspective: Is the Land a Living Entity?

Viewing land not merely as a resource but as an entity possessing rights marks a crucial turning point in environmental philosophy. As Arne Naess emphasized in his theory of Deep Ecology, nature and its resources are not mere instruments for human use; they are equal forms of life and existence — not exclusive to humankind (Naess, 1989). From this standpoint, land is not a commodity that generates economic benefit but a living component of a holistic ecological system.

The concept of Mother Earth (Pachamama), found in many Indigenous worldviews, symbolizes the sacredness attributed to nature. For example, the

Constitution of Bolivia explicitly recognizes nature as a subject of rights — an entity deserving protection and respect for its vital integrity (Gudynas, 2011). This represents the granting of "moral status" to the soil — a notion radical in modern law and philosophy, yet increasingly embraced worldwide.

Terminological Distinctions: Land and Soil

Although "land degradation" and "soil degradation" are sometimes used interchangeably, there are important distinctions between them. "Land" represents a broader ecological system that encompasses soil, vegetation, water, climatic characteristics, and human intervention. "Soil," on the other hand, is the biophysical component of that system — encompassing all levels of biological organization, from molecular to population scale.

According to the UNCCD (1994), land is defined as a "terrestrial bio-productive system that includes soil, vegetation, other biota, and the ecological and hydrological processes operating within it." Thus, not every soil is land, but every land includes soil. This distinction is subtle but critical when formulating policies and designing interventions.

The Epistemology of Understanding Land Degradation

Epistemology seeks to answer questions such as: "What do we know?", "What does it mean to say that we know something?", and "How do we justify our beliefs and observations?" Land degradation often unfolds as an invisible and slow-moving process — one that rarely manifests immediately to the human eye. Hence, understanding land degradation is, in many ways, the art of perceiving the unseen. The types of knowledge applied here go beyond scientific data — they also include local wisdom, traditional observations, and intuitive insight.

While "modern" science quantifies degradation through measurable indicators, traditional societies perceive it through the sound of birds, the scent of the soil, or the timing of leaf fall. This diversity of knowing calls for an epistemological pluralism (Escobar, 1995): to protect the land, no single form of knowledge is sufficient; science, tradition, and ethics must work together.

Global and Local Perspectives: The Anatomy of Degradation in Numbers

The Global State of Land Degradation

Approximately 25% of the Earth's land surface is already significantly degraded (FAO, 2021). According to the Intergovernmental Science-Policy Platform on Biodiversity and Ecosystem Services (IPBES, 2018), this includes about 46% of Africa, 30% of Asia, and 20% of Latin America. Every year, an estimated 12 million hectares of fertile land are lost due to poor agricultural practices, erosion, desertification, and climate change (UNCCD, 2022). This is equivalent to losing an area roughly the size of South Korea each year. By 2050, land degradation is projected to directly affect 1.5 billion people and could force up to 700 million individuals to migrate due to the loss of productive land and livelihood opportunities (World Bank, 2020). Land degradation not only threatens agricultural productivity but also contributes to the deepening of the climate crisis by increasing carbon emissions. Degraded soils release between 3.6 and 4.4 billion tons of CO_2 annually, exacerbating global warming and reducing the planet's natural carbon sequestration capacity (UNEP, 2016).

Furthermore, degradation processes disrupt ecosystem services, reduce water retention capacity, and accelerate biodiversity loss. This degradation of natural capital has profound social and economic consequences, including food insecurity, poverty, and the erosion of cultural ties to the land.

Regional Reflections: The Case of Türkiye and Its Surrounding Geography

Türkiye, located at the intersection of the Mediterranean, Middle Eastern, and Central Asian climatic zones, is among the countries most vulnerable to land degradation and desertification. Nearly 60% of its land area is under the threat of desertification (Ministry of Agriculture and Forestry, Türkiye, 2023). The Konya Closed Basin, Harran Plain, and Central Anatolia regions stand as clear examples of areas where degradation is progressing rapidly — not only due to natural climatic conditions but also because of unsustainable land use practices. In the Konya Basin, for instance, excessive groundwater extraction for irrigation has led to widespread land subsidence and sinkholes, while declining soil organic matter continues to reduce fertility. The Harran Plain, once celebrated for its fertility, now faces salinization, erosion, and structural soil degradation as a result of over-irrigation and mismanagement.

These regional examples illustrate that land degradation is not merely an environmental phenomenon — it is also a developmental, socio-economic, and ethical issue. As the pressure on land continues to grow, the line between ecological limits and human ambition becomes increasingly blurred.

Socio-Economic Dimensions of Degradation

The global economic cost of land degradation is estimated to exceed $6–10 trillion annually, equivalent to 10–15% of global GDP (ELD Initiative, 2015). For many developing countries, this economic loss translates directly into declining agricultural productivity, increased food imports, and deepening rural poverty. In Africa and Asia, where the majority of rural populations depend directly on land for their livelihoods, degradation threatens food sovereignty and social stability. Women and smallholder farmers are disproportionately affected — not only because they rely more heavily on land-based resources, but also because they often lack secure land tenure and access to adaptation financing.

Therefore, land degradation cannot be seen as an isolated environmental problem; it is an interconnected challenge linking ecology, economy, and equity.

Regional Differences and Impacts

a) Sub-Saharan Africa

- More than 3 million hectares of land are degraded every year.
- It is estimated that 300 million people will be displaced by 2030.
- GDP losses may exceed 10% (UNCCD, 2022).

b) South Asia

- In India, land degradation is estimated to cause an annual economic loss of approximately USD 48 billion (FAO, 2021).
- In Pakistan, around 1.5 million hectares of land lose their agricultural productivity every year.

c) Latin America

- Severe degradation occurs especially in the Amazon region, due to deforestation and mining activities.
- The annual economic loss is estimated to be around USD 20 billion.

d) Mediterranean and Europe

- In Spain, Greece, and Italy, the risk of desertification exceeds 30%.
- Across Europe, the annual cost of land degradation is estimated at approximately EUR 85 billion (ELD Initiative, 2015).

The State of Land Degradation in Türkiye

Türkiye, located in the Mediterranean Basin, is significantly affected by drought, erosion, and deforestation. Approximately 59% of its total land area is under the risk of desertification, and this proportion continues to increase.

a) Erosion

- 64% of Turkish soils are exposed to moderate to severe erosion risk.
- Hundreds of millions of tons of soil are displaced by rivers every year (DSİ, 2020).

b) Degradation of Agricultural Lands

- 40% of agricultural lands face productivity losses.
- Major causes include salinization, over-irrigation, and loss of organic matter.

c) Deforestation and Pasture Loss

- Since 1960, pasture areas have decreased by over 60%.
- Non-forest land uses have been steadily increasing.

Quantitative Reflections of Degradation

Continent / Region	Affected Area (million ha)	Annual Economic Loss (USD)
Sub-Saharan Africa	320	65 billion
South Asia	160	30 billion
Latin America	120	20 billion
Middle East & North Africa	90	45 billion
Europe	42	85 billion (EUR)
Türkiye	~20	Data limited; estimated 6–8 billion

Sources: IPBES (2018); UNCCD (2022); FAO (2021); ELD Initiative (2015)

Degradation of Ecosystems

Land degradation disrupts not only physical conditions but also biogeochemical and biological processes — the very cycles that sustain life. For example:

- Reduced carbon storage capacity in soils → increased greenhouse gas emissions.
- Loss of vegetation cover → greater risk of sand and dust storms, floods, and biodiversity loss.
- Decline of soil microorganisms' → reduced self-renewal capacity of the soil.

Although it often appears as a local problem, land degradation is, in reality, a global crisis. Therefore, it must be addressed across both spatial scales (global, regional, national) and systemic levels (nature, economy, society). The data demonstrate that this issue is not only environmental, but also deeply intertwined with social justice, migration, poverty, and economic development.

The Economic Burden and Opportunity Costs of Land Degradation

The global annual economic cost of land degradation is estimated at USD 300 billion (IPBES, 2018). When indirect effects (such as rising food prices, migration, and health expenditures) are included, the cost may reach up to USD 9 trillion (ELD Initiative, 2015). The global GDP in 2023 was approximately USD

106 trillion (IMF, 2024). Thus, USD 9 trillion corresponds to about 8.5% of the global economy. Some country-based comparisons are shown below (note: figures are approximate and nominal; factors such as growth, currency exchange, and inflation may vary):

Country (2023)	**GDP (Trillion USD)**	**Approximate Equivalent of USD 9 Trillion in National Economy**
United States	~27	33% of the U.S. economy
China	~19	47% of China's economy
Japan	~4.2	2.1 times Japan's economy
Germany	~4.4	2 times Germany's economy
India	~4.1	2.2 times India's economy
United Kingdom	~3.3	2.7 times the U.K. economy
France	~3.0	3 times France's economy
Türkiye	~1.3	6.9 times Türkiye's economy
Spain	~1.6	5.6 times Spain's economy
Nigeria	~0.5	18 times Nigeria's economy

Sources: IMF (2024); World Bank (2024); UNCCD (2022); IPBES (2018); ELD Initiative (2015)

The Economic Scale of Land Degradation

This cost exceeds the total economies of many low- and middle-income countries several times over. Addressing the problem is therefore not only an environmental necessity but also an essential condition for global economic stability and sustainable development. The estimated annual sectoral losses caused by land degradation worldwide are presented below (IPBES, 2018; UNEP, 2016):

Sector	**Annual Loss Due to Land Degradation (USD)**
Agriculture	100 billion
Water resources	50 billion
Biodiversity	75 billion
Health	25 billion
Energy	30 billion

Sources: IPBES (2018); UNEP (2016)

Costs of Prevention and Restoration

The average per-hectare costs of preventing and restoring land degradation are shown below (ELD Initiative, 2015). As can be observed, protecting the land is far easier and cheaper than attempting to restore it once degraded (Aydoğdu & Canlı, 2025).

Activity	Average Cost (USD/ha)
Preventive Sustainable Management	20–50
Restoration Efforts	100–500

Source: ELD Initiative (2015)

On the other hand, when evaluated in terms of the economic and social returns of land restoration investments, every USD 1 invested in land restoration generates an economic return between USD 7 and 30 (UNCCD, 2022). By 2030, the rehabilitation of 350 million hectares of degraded land could contribute an estimated USD 1.7 trillion to the global economy (FAO, 2021).

Impacts of Land Degradation on Environment, Society, and Health

Land degradation does not merely signify the loss of soil; it represents a fundamental disturbance of ecological balance, social welfare, public health, and agricultural productivity — challenging the very basis of our relationship with nature and natural resources.

The most direct outcome of land degradation is the weakening of soil capacity due to the deterioration of its physical and chemical structure. Processes such as erosion, loss of organic matter, salinization, and compaction reduce the soil's water-holding capacity and disrupt nutrient cycles. In Türkiye, about 40% of agricultural soils experience productivity losses. In regions such as the Aegean and Southeastern Anatolia, due to intensive farming, over-irrigation, and improper fertilization, the organic matter content of soils has fallen below 1% (Bengisu, 2011; Nif Organic, 2025).

The fragmentation and degradation of natural habitats have devastating impacts on both flora (plants) and fauna (animals). In the Amazon Basin, approximately 137 plant and animal species become extinct each year due to deforestation (WWF, 2021). In Türkiye, habitat fragmentation and land degradation threaten species such as the Anatolian leopard, Mediterranean monk seals, and numerous endemic plant species.

Degraded soils also lose their carbon sink capacity, leading to an increase in atmospheric greenhouse gas emissions. Land degradation accounts for roughly 20% of global carbon emissions (IPCC, 2019). In Türkiye, 8–10 million tons of carbon are displaced from soils each year due to erosion, and a significant portion of this carbon is released into the atmosphere (Lal, 2019).

Land degradation leads to declines in agricultural production, disruption of food chains, and increased rural poverty. According to FAO (2021), by 2050, 4 billion people will face water scarcity, and 800 million people will be at serious risk of food insecurity. In Türkiye, around 74% of water resources are used in agriculture, while irrigation efficiency is only about 45%, creating a vicious cycle that simultaneously degrades both soil and water resources (DSİ, 2022).

Land degradation affects vulnerable rural populations most directly, increasing production losses, income instability, and the risk of social exclusion. In Sub-Saharan Africa, rural poverty rates exceed 70% in areas heavily affected by desertification (UNCCD, 2022). In Türkiye, the income of the agricultural population has decreased by 30% in real terms since 2005 (TÜİK, 2023), with land quality decline being a major contributing factor.

As people lose their lands, they are often forced into migration, creating local, national, and transnational socio-economic challenges. According to World Bank (2020) projections, by 2030, up to 700 million people may be displaced due to land degradation and the climate crisis. In Türkiye, climate- and agriculture-induced migration is most evident in Eastern and Southeastern regions, where seasonal labor migration, rural depopulation, and urban marginalization are becoming increasingly common.

The degradation of soil cover also increases atmospheric dust, leading to more frequent sand and dust storms, which in turn exacerbate respiratory diseases such as asthma and COPD. In the Middle East and North Africa, the frequency and duration of dust storms have risen by 70% over the past 30 years (Middleton, 2017).

Degraded soils also retain more pesticide and fertilizer residues, enabling the transfer of toxic substances into food chains and drinking water. According to WHO (2020), 25% of pesticide-related poisonings occur in low- and middle-income countries. As soil quality declines, the mineral content of food also decreases, contributing to stunted growth, iron deficiency, and vitamin B12 insufficiency, especially among children. Food insecurity, in turn, increases stress and depression risks (FAO, 2021).

There exists a systemic connection between land degradation, environment, society, and human health. When this bond is broken, the consequences are felt not only in nature and natural resources but across all domains of human life. Global evidence and the case of Türkiye reveal that land degradation triggers multidimensional crises. Thus, combating degradation is not merely a matter of protecting nature and resources; it is a moral, economic, social, and political imperative for the continuity of life itself.

Interpretive Summary: A Moral and Systemic Imperative

Land degradation is not merely an environmental issue; it is a mirror reflecting the imbalance between humanity and nature. Every eroded field, every vanished forest, and every displaced community speaks to a deeper rupture — one that challenges our ethical, economic, and existential relationship with the Earth.

When land is lost, so too are the cultures, livelihoods, and memories that it sustains. The erosion of soil becomes the erosion of identity.

The data presented throughout this chapter make it clear: the cost of inaction far exceeds the cost of prevention or restoration. Yet beyond the numbers lies a greater truth — that soil, water, and life form a single, indivisible continuum.

"Protecting one means preserving all."

To fight land degradation, humanity must move beyond reactive measures and adopt systemic transformations that integrate policy, governance, economics, and ethics.

This transformation requires not only investment and innovation but also empathy, humility, and moral vision — a recognition that the Earth is not a resource to be consumed, but a living partner in the story of our existence.

The battle against land degradation is, ultimately, a moral covenant with future generations.

If we fail to protect the soil beneath our feet, we endanger not only ecosystems but also the continuity of civilization itself.

The time to act is not tomorrow, but today — while the Earth still whispers, before it can only remember.

Land Degradation Neutrality Approach: Philosophy, Principles, and Implementation Hierarchy

Origin and Core Objective

Land Degradation Neutrality (LDN) is an environmental vision defined under the United Nations Sustainable Development Goals (SDG 15.3) in 2015. LDN aims to restore degraded lands, ensure the sustainable use of conserved lands, and ultimately maintain healthy and productive land resources (UNCCD, 2015). Its fundamental goal is to "achieve a state of no net land degradation by 2030." In other words, the balance between degraded and restored lands must be maintained.

The concept was first proposed at the Rio+20 Conference (2012) as the goal of "Zero Net Land Degradation in the Fight against Desertification," and later institutionalized by the UNCCD. It is a dual model grounded both in technical measurability and in ecological ethics.

Three Core Principles: According to the UNCCD (2016), the implementation of LDN is based on three fundamental principles:

- Avoidance (Avoid): Protecting healthy land and preventing degradation.
- Sustainable Management (Reduce): Applying effective and eco-friendly land use practices.
- Restoration (Reverse): Rehabilitating degraded lands and reinstating their original functions.

This tripartite model aims not only to sustain ecosystem services but also to generate long-term economic and social benefits.

Principle	Objective	Example of Implementation
Avoidance	Prevent degradation	Limiting land expansion
Sustainability	Reduce intensive use	Agro-ecological farming
Restoration	Reverse degradation	Afforestation and rangeland rehabilitation

Implementation Hierarchy

The LDN approach is not merely a goal—it also represents a systematic implementation process. According to the UNCCD, this process consists of five stages:

- *Establishing a Baseline:* Assessing the current status using indicators such as soil health, carbon stock, and land productivity.
- *Identifying Target Areas:* Selecting priority intervention zones or hotspots.
- *Policy Alignment:* Integrating LDN targets into national policies (e.g., National Strategy to Combat Desertification).
- *Monitoring and Evaluation:* Measuring progress through long-term indicators (e.g., SDG 15.3.1 indicator).
- *Community Participation:* Engaging local people, farmers, women, and youth in the process.

This structure is grounded not only in technical expertise but also in moral responsibility: the future of the soil is shaped by the decisions we make today.

Philosophical and Ethical Foundations

LDN is not merely about "saving land"; it is a call to transform humanity's relationship with the environment, nature, and natural resources. This call incorporates key elements of ecological ethics and environmental philosophy:

- Arne Naess's Deep Ecology advocates viewing soil as a moral subject rather than an object.
- Hans Jonas's Ethics of Responsibility emphasizes the duty of current generations toward future ones.
- Bruno Latour's Politics of the Planet proposes seeing nature and resources not as objects to be protected but as coexisting actors in a shared world.

LDN has the potential to translate these philosophical approaches into practical action. Every piece of land is not merely an economic asset, but a fragment of the future.

Türkiye and Land Degradation Neutrality: Target Process and Challenges

Türkiye's struggle against land degradation began with erosion control programs in the 1960s; however, these efforts remained fragmented and sectoral for decades. After the adoption

of the UNCCD in 1994, Türkiye became a party to the Convention and prepared its first National Action Program to Combat Desertification (NEAP) in 2004. Yet, these frameworks were largely principle-based and lacked operational focus.

In 2015, during the 12th Conference of the Parties (COP12) held in Ankara, Türkiye officially adopted the LDN target and established a national policy framework aligned with SDG 15.3's "Zero Net Land Degradation" vision. This process restructured institutional frameworks, legal mechanisms, and implementation tools, setting national targets within a 2030 vision (UNCCD, 2015).

Türkiye's national LDN targets were built upon afforestation, rangeland rehabilitation, erosion control, and soil conservation practices in dryland farming. Three key national documents stand out in this process:

- National Strategy and Action Plan to Combat Desertification
- National LDN Target Report
- National Monitoring Indicators Harmonized with Agricultural, Water, Forest, and Soil Policies

In these documents, Türkiye committed to addressing land degradation not merely as an issue of drought or desertification but as an ecosystem-based, cross-sectoral, and human-centered threat.

Türkiye's Land Degradation Neutrality Target Process

1. 2015 – Initiation

- Türkiye officially adopted the vision of Land Degradation Neutrality during the UNCCD COP12 in Ankara.
- The same year, it joined the UNCCD "LDN Target Setting Program (LDN-TSP)."
- Thus, Türkiye committed to establishing national targets to achieve LDN by 2030 (UNCCD, 2015).

2. 2016–2017 – Initial Target Setting

- In 2016, Türkiye and the UNCCD Secretariat jointly organized the "National LDN Target Setting Workshop."
- In 2017, Türkiye reported its official national LDN targets to the UNCCD, including:
    - 1 million hectares of afforestation,
    - 750,000 hectares of rangeland rehabilitation,
    - 2,2million hectares of soil and water conservation in dryland farming areas,
    - Increasing forest cover to 30% of the national territory (UNCCD, 2017).

3. 2018–2021 – Monitoring and Interim Reports

- Türkiye began integrating its LDN targets into the UNCCD's monitoring system (PRAIS3, later PRAIS4).
- Under SDG 15.3.1, the national land degradation rate decreased from 14.3% to 13.4% between 2019 and 2021 (United Nations, 2021).
- However, challenges such as water scarcity, salinization, and rapid urbanization posed sustainability risks for LDN implementation.

4. 2022 – Update and Revision

- In 2022, Türkiye submitted an updated national report via the UNCCD PRAIS4 system.

This report maintained previous targets while:

- Shifting monitoring methodologies toward watershed-based and indicator-driven models,
- Highlighting the linkages between climate change and land degradation (UNCCD, 2022).

5. 2024–2030 – New Strategy

- In 2024, the General Directorate of Combating Desertification and Erosion published the National Strategy and Action Plan to Combat Desertification (2024–2030).

Updated LDN goals include:

- Basin-based land management,
- Irrigation efficiency,
- Forest fire resilience,
- Rehabilitation of degraded lands.

Legislative and Institutional Framework

Türkiye's land protection policies are administered through numerous laws, regulations, and administrative bodies, sometimes leading to jurisdictional overlaps.

a) Key Legislative Documents:

- Soil Conservation and Land Use Law No. 5403 (2005)
- Pasture Law No. 4342 (1998)
- Forest Law No. 6831 (1956)
- Organic Agriculture Law No. 5262 (2004)
- Strategy Document for Combating Desertification (2019–2030)
- National Strategy and Action Plan to Combat Desertification (2024–2030)

These laws regulate the protection of soils, prevention of non-agricultural land use, pasture rehabilitation, and supervision of agricultural production.

b) Key Responsible Institutions:

- Ministry of Environment, Urbanization and Climate Change
- Ministry of Agriculture and Forestry
- General Directorate of Combating Desertification and Erosion (ÇEM)
- General Directorate of Forestry (OGM)
- State Hydraulic Works (DSİ)
- General Directorate of Water Management (SYGM)
- General Directorate of Agricultural Reform (TRGM)
- General Directorate of Agricultural Research and Policies (TAGEM)
- General Directorate of Plant Production (BÜGEM)
- Local Governments (Provincial/District Directorates, Municipalities)

However, improvements are still needed in data sharing, strategic coordination, and field-level implementation among these institutions.

Strengths of Türkiye's LDN Framework

- *Institutional Capacity:* Agencies such as ÇEM, OGM, DKMP, and TAGEM actively operate within their fields.
- *Strategic Action Plans:* A clear roadmap for LDN implementation exists.
- *International Alignment:* Türkiye is among the few countries to have fully adapted UNCCD goals to its national context.
- *High Afforestation Capacity:* During the period between 2008 and 2022, 5.5 billion saplings were planted, making Türkiye first in Europe and fourth globally in afforestation.
- *Long-Term Institutional Experience:* Active participation in the UNCCD process since 1994.
- *Data Infrastructure:* Advanced satellite monitoring and GIS-based national land observation systems established.

Challenges

Key challenges faced by Türkiye in LDN implementation include:

- *Institutional Fragmentation:* Limited data and authority sharing among institutions (ÇEM, DSİ, OGM, DKMP, SYGM, TAGEM, Municipalities, etc.).
- *Sectoral Misalignment:* Water, energy, agriculture, and urbanization policies are not always consistent with LDN objectives.
- *Limited Public Awareness and Participation:* Awareness among rural communities, farmers, NGOs, and local governments remains low; education and incentive mechanisms are insufficient.
- *Financing Constraints:* LDN projects largely depend on external funds (GEF, UNDP), while domestic funding remains limited.
- *Excessive Land Use Pressure:* Industrial, mining, and urban expansion continues to exert high pressure on land.
- *Data Gaps and Transparency Issues:* Lack of updated and accessible data for several indicators.
- *Coordination Deficits:* Cross-institutional data sharing and methodological harmonization for SDG 15.3.1 are still incomplete.
- *Drought and Water Scarcity:* Groundwater depletion poses severe threats, especially in the Konya Closed Basin.
- *Salinization and Sodicity:* In areas like the Harran Plain, inadequate drainage and poor irrigation practices increase soil salinity.
- *Loss of Productive Soils*: Rapid urbanization and land conversion reduce agricultural land.
- *Climate Risks:* Increasing forest fires after 2021 threaten the resilience and quality of forest ecosystems despite quantitative gains (ÇEM, 2022).

Implementation Programs Related to LDN in Türkiye

Program	Geographic Scope	Scope and Impact
Afforestation and Erosion Control Action Plan	Nationwide	6.5 million ha afforested between 2008–2023
Rangeland Rehabilitation and Management Project	Eastern and Central Anatolia	2.1 million ha rehabilitated by 2023

Soil Organic Matter Enhancement Project	Aegean, Central Anatolia	Increased incentives for organic and green manuring
National Water Efficiency Action Plan	GAP, Konya, Thrace	500,000 ha transitioned to drip irrigation after 2023
Micro-Watershed Rehabilitation Projects	Pilot sites in 81 provinces	Integrated soil and water management with community participation

Türkiye has established an ambitious policy framework for achieving land degradation neutrality. However, institutional, financial, and social barriers hinder full realization of these targets. Strengthening multi-stakeholder, bottom-up participatory models and domestic financing mechanisms is therefore essential.

While Türkiye's fight against land degradation rests on strong political will and technical investment, the success of LDN ultimately depends on extending this awareness to the societal level—ensuring that local actors participate actively and the public reconnects with the land. The success of LDN will not be achieved solely through government programs but through the people's renewed relationship with the soil.

International Comparison of Countries Setting LDN Targets: Where Does Türkiye Stand?

Country	**LDN Implementation Status**	**Local Participation**	**Data Monitoring System**	**Resources**
Türkiye	Moderate–Advanced	Weak–Moderate	Advanced	Mixed–Insufficient domestic financing
Spain	Advanced	Strong	Advanced	State-supported
Germany	Advanced	Strong	Integrated	State-supported
Senegal	Moderate	Strong	Insufficient	External fund-oriented
India	Moderate	Moderate	Developing	Mixed
Niger	Limited	Weak	Insufficient	External fund-oriented

Türkiye stands ahead of many countries in terms of technical capacity and strategic planning; however, it lags behind developed countries such as Germany and Spain in public participation, transparency, practical implementation, and local capacity building. While Türkiye possesses an advanced technical infrastructure, it occupies a limited position regarding participation and coordination — a condition that directly affects the success potential of LDN targets.

The LDN approach represents a paradigm shift in the relationship between nature and humanity. In this model, the environment, nature, and natural resources are no longer commodities to be consumed but entities with which we must coexist. Balance can be achieved not merely by bringing land degradation to zero but by establishing a moral contract with nature.

Türkiye possesses significant institutional infrastructure and large-scale projects under the LDN framework; however, the water–soil nexus remains the most fragile domain. Türkiye demonstrates notable progress, particularly in expanding forest areas and achieving partial improvements in the proportion of degraded lands. Nevertheless, water scarcity, salinization, and climate change impacts remain critical threats to the sustainability of land degradation neutrality. Achieving the 2030 targets depends on strengthening water management, land-use planning, and effective local-level monitoring mechanisms.

Evolution of the LDN Concept in Light of United Nations Models: 1994 – 2007 – 2015

The United Nations brought the issue of land degradation to the global agenda for the first time at the 1992 Rio Earth Summit, leading to the adoption of the United Nations Convention to Combat Desertification (UNCCD) in 1994. Over time, this process evolved beyond the focus on desertification into an ecosystem-based, sustainability-oriented, multi-level, and equity-centered approach. This evolution can be traced through three critical milestones (Aydınoğlu & Aydoğdu, 2025):

1994: Institutional Beginning – The Convention to Combat Desertification

Adoption of the UNCCD: Established in 1994, the UNCCD aimed to institutionalize international solidarity against desertification, particularly in arid, semi-arid, and dry sub-humid regions, with an emphasis on Africa (UNCCD,

1994). During this period, land degradation was mainly addressed within the axis of desertification, and solutions were largely regional in nature.

Main Focus: Soil erosion, climate impacts, drought, migration, and rural poverty. Strategies concentrated on infrastructure development, afforestation, and the establishment of monitoring systems.

Weaknesses:

- Insufficient focus on socio-economic drivers
- Lack of an ecosystem services perspective
- Limited inclusion of local knowledge systems

Philosophical Interpretation:

This period was characterized by a modernist perspective that viewed soil as a resource to be controlled and managed — nature as an object of domination.

2007: The Conference of the Parties (COP8) – The Introduction of LDN

Turning Point: At the 2007 UNCCD COP8 Conference in Madrid, Spain, the concept of land degradation expanded beyond desertification, and the term "Land Degradation Neutrality (LDN)" appeared for the first time in official documents.

New Approach:

- The goal extended beyond prevention to include restoration.
- Quantifiable indicators such as carbon stock, water management, and soil productivity gained prominence.
- Nature-based solutions entered the policy discourse for the first time.

Policy Transformation:

The focus shifted from "reducing degradation" to "balancing degradation with restoration" — marking an ethical paradigm shift centered on equilibrium.

Philosophical Interpretation:

The post-2007 era strengthened the notion of viewing nature not merely as something to be protected but as a self-healing living entity. It represents the first convergence between ecological ethics and policy.

2015: Institutionalization through the Sustainable Development Goals (SDGs)

Culmination of the Transformation: With the adoption of the 2030 Agenda for Sustainable Development in 2015, the LDN approach became one of the global targets:

SDG 15.3: "By 2030, achieve a land degradation-neutral world."

Innovative Features:

- A global commitment encompassing all countries.
- The "no net loss" principle emphasizing zero net degradation.
- Integration of social justice dimensions such as community participation, local knowledge, and women's representation.

Governance Integrity:

- *Monitoring indicators:* Soil productivity, carbon content, land cover change.
- *Policy coherence:* Water management, agriculture, climate adaptation, and biodiversity.

Philosophical Depth:

The post-2015 period institutionalizes the moral necessity of living in harmony with nature — marking the decline of the modern anthropocentric paradigm and recognizing nature as an ethical subject.

Comparative Evaluation: 1994 – 2007 – 2015

Year	Main Approach	Conceptual Focus	Implementation Dimension	Ethical Perspective
1994	Technical–protection-oriented	Desertification	Afforestation, monitoring	Nature = resource
2007	Systemic	Land health	Degradation–restoration balance	Nature = ecological actor
2015	Holistic	Sustainability and Land Rights	LDN – policy alignment & participation	Nature = rights-bearing entity

The United Nations' perspective on land degradation has evolved from technical intervention to ethical transformation. While the 1994 framework reflected a mindset of "fighting" against nature, the post-2015 paradigm emphasizes forming an alliance with nature. This evolution signifies not only a shift in policy frameworks but also a profound moral and philosophical deepening in humanity's relationship with the Earth.

Land Rights, Justice, and Ecological Ethics: A Philosophical Perspective

Is It Enough to See Land Only as a Resource?

The modern capitalist system has largely commodified nature: land has become an object of use for agriculture, mining, urbanization, and energy production. This worldview is based on an anthropocentric perspective — nature and natural resources exist for human use. This mode of thought has become both the cause and the consequence of today's land degradation (Plumwood, 1993).

From a philosophical standpoint, a critical question arises: *Is land merely an instrument, or is it a moral subject?*

Land Rights and the Subjectivity of Nature

The concept of "land rights" should not be defined solely through ownership, but through the right of the environment, nature, and natural resources to exist, regenerate, and produce life.

a) Deep Ecology (Naess, 1973):

All living beings and elements of nature must be protected not merely for their utility but for their intrinsic value. Soil, water, and air are ethical interlocutors, not passive entities.

b) Rights-Based Approach:

The Indigenous concept of Mother Earth recognizes nature as both a legal and moral subject. In Bolivia and Ecuador, constitutions have granted nature the status of a "rights-bearing entity" (Gudynas, 2011).

c) New Materialism and Eco-socialism:

These approaches view nature not as a passive "environment" but as an interactive actor engaged in reciprocal relationships with humanity. They challenge the binary separation between human and nature.

Ecological Justice and Land

Ecological justice encompasses not only the equitable distribution of resources among humans but also the ethical regulation of relationships between humans, resources, and the natural world. Land degradation violates this justice in several ways:

- *Intergenerational Justice:* Today's degradation infringes upon the right to life of future generations (Jonas, 1984).
- *Geographical Injustice:* Some regions — particularly those in the Global South — are directly affected, while the primary causes of degradation stem from Northern countries.
- *Social Justice:* Vulnerable groups — such as peasants, migrants, women, the elderly, and children — are disproportionately affected by land degradation.

In the case of Türkiye, the Southeastern, Eastern Anatolia, and Central Anatolia regions are the most fragile in terms of both natural resources and socio-economic structures. Agricultural policies remain insufficient, and small-scale farmers are increasingly disconnected from their land — a condition that intertwines land rights with social inequality (TÜİK, 2023).

Is an Ethical Contract with Nature Possible?

Moral responsibility should not exist solely among humans but also between humans and nature, including its resources. The following ethical approaches offer guidance:

a) Hans Jonas' Ethics of Responsibility:

"As the power to act increases, so too must the scope of responsibility."

In other words, as technological interventions amplify the harm inflicted upon nature and natural resources, our moral, ethical, and financial obligations must proportionally expand (Jonas, 1984).

b) Latour's Planetary Politics:

The environment, nature, and natural resources are no longer mere backdrops to human activity but central political and social actors. Ignoring the voice of the planet is not only unethical — it is existential blindness (Latour, 2017).

c) "Soil Justice" Approach:

Soil is the spatial foundation of justice. Through its ecological functions — ecosystem services, food production, and water cycling — it ensures the livelihood security of society's most vulnerable groups. Therefore, the degradation of soil equates to the erosion of justice itself (Berry, 2010).

Reflections in Policy and Law

Land-related policies remain largely property-based and production-oriented. Viewing nature and natural resources as ethical actors requires fundamental transformations:

- The Right to Land should be constitutionally guaranteed.
- The legal personhood of nature — including rivers, mountains, and forests — should be recognized.
- Agricultural policies should aim not only at productivity but also at ecological integrity and social justice.

- Local communities and peasants should be acknowledged as legal representatives of nature.

Land degradation is not merely an environmental or resource problem; it is also a matter of justice, ethics, and rights. The soil is not just the ground upon which we live — it is a living entity with its own right to existence. Protecting it is not only a responsibility but also a moral imperative.

The new contract with nature must not be economic in nature but existential — an agreement grounded in reconciliation and protection of the soil. Only through such an ethical covenant can humanity restore balance with the Earth.

Sustainable Solutions and New Policy Directions: Adaptation, Restoration, and Participation

The success of combating land degradation depends not only on technical measures but on holistic, multi-level strategies that redefine the relationship between humans and nature. In this context, three key concepts must be placed at the core of the solution: adaptation, restoration, and participation.

Adaptation: Enhancing the Resilience of Society and Ecosystems

Ecological adaptation refers to the land's capacity to maintain its resistance against stress factors such as climate, land-use pressure, and natural disasters. Social adaptation, on the other hand, is the ability of local communities and farmers to withstand these changes socio-economically.

Some Policy Recommendations:

- *Climate-Sensitive Land Use:* Crop patterns in agricultural production should be restructured according to rainfall regimes and drought risks.
- *Soil Health Monitoring:* Soil health maps should be developed for each region, measuring soil performance and providing region- and farmer-specific recommendations for fertilization, irrigation, and cropping.

- *Rural Early Warning Systems:* Sensor-based, community-centered early warning networks should be established and expanded to monitor droughts, floods, and landslides.

Implementation Example:

Senegal's "Great Green Wall" project has protected land and strengthened rural livelihoods through agroforestry practices implemented jointly with local communities in desertified areas.

Restoration: A Process of Reconciliation with the Soil

Restoration is not merely about making degraded soils "usable" again—it is a process of reestablishing a relationship with nature and resources. In this process, not only ecosystems but also social structures must be rehabilitated.

Some Policy Recommendations:

- *Ecosystem-Based Restoration:* Afforestation with native species, rangeland rehabilitation, and wetland recovery projects should be further expanded.
- *Incentives for Carbon Farming:* Practices that enhance carbon sequestration in soils (such as cover crops, no-till farming, and green manuring) should be publicly supported.
- *Restoration Credits:* Public-private partnerships should invest in land rehabilitation projects, offering farmers incentives based on soil health outcomes.

Implementation Example:

China's Loess Plateau Restoration Program restored productivity and water balance across 3 million hectares in 20 years and contributed to lifting 2.5 million people out of poverty (World Bank, 2007).

Participation: Engaging Society in the Process

Achieving sustainable solutions and lasting success is only possible when farmers, villagers, women, youth, civil society organizations, academia, and local governments act together—not through state action alone. Participatory models are, therefore, a strategic necessity.

Some Policy Recommendations:

- *Local Land Councils:* Effective decision-making platforms should be established in every province and district, involving farmers, NGOs, municipalities, and experts.
- *Soil Ethics Education:* From primary school to university, curricula should be restructured to focus on nature and soil rights.
- *Empowering Women:* Women should be supported as decision-makers in land use, cooperatives, and rural governance processes.

Implementation Example:

India's Watershed Plus program, in which women lead land restoration and rural development initiatives, has both improved land productivity and strengthened social resilience.

An Integrated Roadmap for Türkiye

- *Nature-Based Soil Legislation:* Land rights should be legally reinforced, and nature should be recognized as a subject of rights.
- *National LDN Legislative Reform:* Laws No. 5403 and 4342 should be revised to better align with the principles of Land Degradation Neutrality (LDN).
- *Investment and Financial Mechanisms:* Agricultural subsidies should be redesigned around soil health and supported by green financing tools.
- *Data Transparency and Open Access:* Land use, carbon, and productivity data should be made more publicly accessible.

- *Participation from Local to National:* Micro-watershed plans developed at the village level should be more effectively integrated into national strategies.

A sustainable future is possible not only by protecting nature but also by recognizing the story, culture, and rights of the soil. Adaptation, restoration, and participation constitute a three-pillar life strategy against land degradation—a strategy that defends not only nature but also humanity itself.

Chapter Conclusion

Soil cannot speak—but it screams in silence. Through the slow erosion of its fertile layers, through the cracks of its thirsty skin, through the sorrow of its diminishing fertility… As soil disappears, what is lost is not merely a physical surface, but also a memory, a culture, and a way of life.

In this sense, land degradation does not simply mean environmental deterioration—it signifies a breach in the covenant between humanity and nature. Desertified lands, displaced rural communities, vanishing biodiversity, and depleted resources… All of these are warnings, calls, and signals for awakening.

This is not merely an issue of environment, nature, or natural resources—it is a question of humanity and the future. For it is we who degrade the soil, and only we can restore it?

The "modern world" promised soil productivity, growth, and development—but gave it drought, chemicals, concrete, and exhaustion. Modernity measured productivity solely in economic gain, disregarding nature's silent productivity, cyclicality, and compassion. The UN's goal of Land Degradation Neutrality is a delayed but valuable attempt to heal this rupture. Yet technical interventions alone are insufficient. What we truly need is a new moral pact between humans and nature.

We Must Now Ask Ourselves:

- What kind of relationship do we build with the soil—ownership or companionship?
- Does the concept of "sustainability" in policy documents truly reflect the voice of local people?

- Can development occur at the expense of the soil, or is prosperity possible with the soil?

If we cannot answer these questions honestly, we will lose not only the soil but also our history, culture, and future. A new path, a new understanding, is possible. The ultimate truth is this: Soil is not merely the ground beneath our feet. It is us. We are its continuation.

"It is GOD Who created you from dust, then from a drop of fluid, then from a clot." (Qur'an, Al-Mu'min 40:67)

And as Mevlânâ beautifully reminds us: "We came from the soil and to the soil we shall return; what matters is not to turn into mud."

Thus, sustainability is not merely a technical goal—it is an ethical responsibility, a cultural reconnection, a vital act of resistance. When humanity learns to live with nature, not against it, then and only then will development, justice, and peace become truly possible.

O human being, remember the environment, nature, and resources you have forgotten in the city. Touch the soil lost between the concrete walls. Take a handful of earth in your hands. Within it, you will find the labor of millennia, the friendship of rain and sun, the calluses of farmers, and the poetry of nature. Our answer to the silent cry of the soil must be this: To live again—together—with justice, knowledge, ethics, philosophy, and love.

Chapter References

Akın, T. & Berberoğlu, S. (2024). Arazi Tahribatı Çerçevesinde Toprak Organik Karbon Stoklarındaki Değişimin Konumsal Bilgi Sistemleriyle Modellenmesi *(Modeling the Change in Soil Organic Carbon Stocks within the Framework of Land Degradation Using Geographical Information Systems).* Retrieved from https://www.researchgate.net/publication/385815164_ARAZI_TAHRIBATI_CERCEVESINDE_TOPRAK_ORGANIK_KARBON_STOKLARINDAKI_DEGISIMIN_KONUMSAL_BILGI_SISTEMLERIYLE_MODELLENMESI

Altürk, B., Akgöz, R., Sener, M. & Konukcu, F. (2025). İklim Değişikliği-Arazi Tahribatı ve Arazi Tahribatının Modellenmesi *(Climate Change–Land Degradation and Modeling of Land Degradation).* Retrieved from https://www.researchgate.net/publication/392746108_IKLIM_DEGISIKLIGIA RAZI_TAHRIBATI_VE_ARAZI_TAHRIBATININ_MODELLENMESI

Aydınoğlu, E. & Aydoğdu, M. H. (2025). Arazi Bozulumunu Dengeleme Mücadelesi Süreci: Kumaş Aynı Modeller Farklı *(The Process of Combating Land Degradation Neutrality: Same Fabric, Different Models).* ICHEAS 9th International Conference on Health, Engineering and Applied Sciences, March 14–16, Casablanca–Morocco, pp. 82–93. ISBN: 978-625-5962-51-5

Aydoğdu, M. H. & Canlı, P. (2025). Arazi Tahribatı: Küresel Bir Tehdit Olarak Önemi, Etkileri ve Sürdürülebilir Çözüm Yaklaşımları *(Land Degradation: Its Importance, Impacts, and Sustainable Solution Approaches as a Global Threat).* Latin America 10th International Conference on Scientific Researches, May 1–4, Havana/Cuba, pp. 435–442. ISBN: 978-625-5962-70-6

Aytop, H. (2024). CORINE arazi örtüsü/arazi kullanım sınıflarına göre Türkiye'deki arazi sınıflarının zamansal ve mekânsal değişimi *(Temporal and Spatial Change of Land Classes in Turkey According to CORINE Land Cover/Land Use Classes). Toprak Bilimi ve Bitki Besleme Dergisi (Journal of Soil Science and Plant Nutrition)*, DOI: 10.33409/tbbbd.1491229

Bengisu, G. (2011). GAP Bölgesinde Sürdürülebilir Tarım İçin Ekim Nöbeti Sistemleri *(Crop Rotation Systems for Sustainable Agriculture in the GAP Region). Alınteri*, 20(B), 33–39. Retrieved from https://dergipark.org.tr/tr/download/article-file/26253

Berry, W. (2010). *The Art of the Commonplace: The Agrarian Essays of Wendell Berry.* Counterpoint. Retrieved from https://www.amazon.com/Art-Commonplace-Agrarian-Essays-Wendell/dp/1593760078

Bilgili, A. V., Küçük, Ç. & Van Es, H. M. (2017). *Assessment of the quality of the Harran Plain soils under long-term cultivation. Environmental Monitoring and Assessment*, 189, 460. DOI: 10.1007/s10661-017-6177-y

Bulut, F. Ş., Ateşoğlu, A. & Acar, E. (2025). Türkiye İçin Küresel Arazi Kullanımı/Örtüsü Haritalarının Doğruluğunun Değerlendirilmesi: ESA World Cover ve ESRI Land Cover Karşılaştırması *(Evaluation of the Accuracy of Global Land Use/Cover Maps for Turkey: Comparison of ESA World Cover and ESRI Land Cover). Bartın Orman Fakültesi Dergisi (Bartin Faculty of Forestry Journal)*, 27(2), 223–238. DOI: 10.24011/barofd.1666812

Çelik, A. (Ed.) (2024). *Management of Soil and Plant Health in the Era of Climate Change.* İksad Publication, Ankara. ISBN: 978-625-378-048

ÇEM (2019). Türkiye Çölleşme ile Mücadele Raporu *(Türkiye Desertification Combat Report).* Ankara: Çölleşme ve Erozyonla Mücadele Genel Müdürlüğü. Retrieved from https://cem.csb.gov.tr/

ÇEM (2022). Ulusal Çölleşmeyle Mücadele Stratejisi ve Eylem Planı (2024–2030) *(National Strategy and Action Plan to Combat Desertification (2024–2030)).* Ankara: Çölleşme ve Erozyonla Mücadele Genel Müdürlüğü. Retrieved from https://cem.csb.gov.tr/

ELD Initiative. (2015). *The value of land: Prosperous lands and positive rewards through sustainable land management.* Economics of Land Degradation Initiative. Retrieved from https://www.eld-initiative.org

Escobar, A. (1995). *Encountering development: The making and unmaking of the Third World.* Princeton University Press. Retrieved from https://www.amazon.com/Encountering-Development-Making-Unmaking-Third/dp/0691150451

FAO & UNEP. (2021). *The State of the World's Land and Water Resources for Food and Agriculture – Systems at breaking point.* Retrieved from https://www.fao.org/3/cb7654en/cb7654en.pdf

FAO. (2021). *Land Degradation and Sustainable Agriculture in Europe and Central Asia.* Retrieved from https://www.fao.org/

GEF. (2020). *Financing Sustainable Land Management.* Retrieved from https://www.thegef.org

Gudynas, E. (2011). *Buen Vivir: Today's Tomorrow. Development,* 54(4), 441–447. https://doi.org/10.1057/dev.2011.86

IMF. (2024). *International Monetary Fund, World Economic Outlook Database.* Washington, DC: IMF.

IPBES. (2018). *The Assessment Report on Land Degradation and Restoration.* Intergovernmental Science-Policy Platform on Biodiversity and Ecosystem Services. Retrieved from https://ipbes.net

IPBES. (2018). *Intergovernmental Science-Policy Platform on Biodiversity and Ecosystem Services Land Degradation and Restoration Assessment.* Bonn: IPBES. Retrieved from https://ipbes.net

IPCC. (2019). *Climate Change and Land – Summary for Policymakers.* Retrieved from https://www.ipcc.ch

İmamoğlu, A. (2025). Farklı Arazi Kullanımlarında Toprak Organik Karbon Miktarının Belirlenmesi: Alaçam İlçesi Örneği *(Determination of Soil Organic Carbon Amount under Different Land Uses: The Case of Alaçam District). Black Sea Journal of Engineering and Science*, 8(5), 1617–1627. DOI: 10.34248/bsengineering.1739969

Jonas, H. (1984). *The Imperative of Responsibility: In Search of an Ethics for the Technological Age.* University of Chicago Press, US.

Kimmerer, R. W. (2013). *Braiding Sweetgrass: Indigenous Wisdom, Scientific Knowledge and the Teachings of Plants.* Milkweed Editions. Retrieved from https://www.amazon.com/Braiding-Sweetgrass-Indigenous-Scientific-Knowledge/dp/1571313567

Kocalar, A. C. (2025). *Water Scarcity – Türkiye - Konya and Niğde. International Journal of Water Management and Diplomacy.* Retrieved from https://dergipark.org.tr/en/download/article-file/4420279

Lal, R. (2019). *Accelerated Soil Erosion as a Source of Atmospheric CO_2. Soil and Tillage Research*, 188, 35–40.

Latour, B. (2017). *Facing Gaia: Eight Lectures on the New Climatic Regime.* Polity Press. Retrieved from https://grattoncourses.wordpress.com/wp-content/uploads/2019/06/bruno-latour-facing-gaia-eight-lectures-on-the-new-climatic-regime.pdf

Middleton, N. (2017). *Desert dust hazards: A global review. Aeolian Research*, 24, 53–63.

Naess, A. (1973). *The Shallow and the Deep, Long-Range Ecology Movements. Inquiry*, 16(1), 95–100.

Naess, A. (1989). *Ecology, community and lifestyle: Outline of an ecosophy.* Cambridge University Press, UK.

Nif Organic. (2025). Organik Madde Toprağın Canıdır *(Organic Matter Is the Soul of the Soil).* Retrieved from https://niforganik.com.tr/organik-madde-topragin-canidir/

OGM. (2023). 2023 İdare Faaliyet Raporu *(2023 Administrative Activity Report).* Ankara: Orman Genel Müdürlüğü (General Directorate of Forestry). Retrieved from https://www.ogm.gov.tr/tr

Ostrom, E. (2009). *Governing the Commons: The Evolution of Institutions for Collective Action.* Cambridge University Press, UK.

Plumwood, V. (1993). *Feminism and the Mastery of Nature.* Routledge. Retrieved from https://www.routledge.com/Feminism-and-the-Mastery-of-Nature/Plumwood/p/book/9780415068109

Shiva, V. (2016). *Who Really Feeds the World?.* North Atlantic Books.

SYGM. (2022). Değişen İklime Uyum Çerçevesinde Su Verimliliği Strateji Belgesi (2023–2033) ve Eylem Planı *(Water Efficiency Strategy Document (2023–2033) and Action Plan within the Framework of Adaptation to the Changing Climate).* Ankara: Su Yönetimi Genel Müdürlüğü (General Directorate of Water Management). Retrieved from https://suverimliligi.gov.tr/wp-content/uploads/2023/08/SU-VERIMLILIGI-STRATEJI-BELGESI-ve-EYLEM-PLANI-2023-2033.pdf

T.C. Çevre, Şehircilik ve İklim Değişikliği Bakanlığı. (2023). İklim Değişikliği Uyum Eylem Planı *(Climate Change Adaptation Action Plan). Retrieved from* https://www.csb.gov.tr/

Temesgen, T. (2018). *Review of Saline Water Irrigation Logging and Salt Affected Soil on Maize (Zea Mays L.) Yield and Management's. International Journal of Food Science and Agriculture*, 2(5), 95–107. DOI: 10.26855/ijfsa.2018.06.001

Tunç, G., & Kaman, H. (2022). *Salinity Problem in Agricultural Lands. Turkish Journal of Agriculture, Food Science and Technology*, 10(11), 2232–2236. https://doi.org/10.24925/turjaf.v10i11.2232-2236.5549

TÜİK. (2022). Tarım Alanları ve Su Kullanımı İstatistikleri *(Agricultural Land and Water Use Statistics). Retrieved from* https://www.tuik.gov.tr/

TÜİK. (2023). Tarımsal Gelir ve Arazi Kullanımı İstatistikleri *(Agricultural Income and Land Use Statistics).* Retrieved from https://www.tuik.gov.tr/

UNCCD. (1994). *United Nations Convention to Combat Desertification.* Retrieved from https://www.unccd.int

UNCCD. (2007). *COP8 Decisions and Reports.* Retrieved from https://www.unccd.int

UNCCD. (2015). *Achieving Land Degradation Neutrality. The Science-Policy Interface.* Retrieved from https://www.unccd.int/publications

UNCCD. (2015). *Report of the Conference of the Parties on its twelfth session (COP12, Ankara).* UNCCD Secretariat. Retrieved from https://www.unccd.int

UNCCD. (2016). *LDN Target Setting Programme Guidance Document.* Retrieved from https://www.unccd.int

UNCCD. (2017). *Land Degradation Neutrality Target Setting Programme: Turkey National Report.* UNCCD Secretariat. Retrieved from https://www.unccd.int

UNCCD. (2022). *Turkey National Report 2022 (PRAIS4).* UNCCD Secretariat. Retrieved from https://www.unccd.int

UNCCD. (2022). *United Nations Convention to Combat Desertification Global Land Outlook, Second Edition.* Bonn: UNCCD. Retrieved from https://www.unccd.int

UNDP India. (2022). *Watershed Plus Final Report.* Retrieved from https://www.undp.org/

UNDP Türkiye. (2023). ATD Projeleri Değerlendirme Raporu *(Evaluation Report of ATD Projects).* Retrieved from https://www.undp.org/

UNEP. (2016). *Global Environmental Outlook GEO-6: Regional Assessments.* United Nations Environment Programme. Retrieved from https://www.unep.org

UNEP. (2016). *Healthy Environment, Healthy People.* Retrieved from https://www.unep.org

UNEP. (2020). *A Practical Guide to Land Degradation Neutrality.* Retrieved from https://www.unep.org

UNEP. (2020). *Land Restoration for Achieving SDG 15.3.* Retrieved from https://www.unep.org

United Nations. (2015). *Transforming our world: The 2030 Agenda for Sustainable Development.* Retrieved from https://sdgs.un.org/2030agenda

United Nations. (2021). *SDG Indicator Metadata: 15.3.1 – Proportion of degraded land over total land area.* UN Statistics Division. Retrieved from https://unstats.un.org/sdgs/metadata/files/Metadata-15-03-01.pdf

WHO. (2020). *Public Health Impact of Chemicals: Known and unknowns.* Retrieved from https://www.who.int

World Bank. (2007). *Loess Plateau Watershed Rehabilitation Project Review.* Retrieved from https://documents.worldbank.org/en/publication/documents-reports/documentdetail/142661468762366534

World Bank. (2020). *Groundswell Report: Preparing for Internal Climate Migration.* Retrieved from https://www.worldbank.org

World Bank. (2024). *World Development Indicators.* Washington, DC: World Bank. Retrieved from https://www.worldbank.org

WWF. (2021). *Living Planet Report 2020.* Retrieved from https://www.wwf.org

8th CHAPTER

SAND AND DUST STORMS: THE SILENT REBELLION OF NATURE

Chapter Summary

In this chapter, sand and dust storms (SDS) are discussed not merely as meteorological phenomena but as a form of response from nature to the imbalance caused by human interference. Human-induced pressures such as ecosystem degradation, climate change, improper land use, and desertification have increased the frequency and intensity of SDS events. While the chapter examines the physical, ecological, socioeconomic, and health impacts of SDS with data-based analyses, it also emphasizes the moral and philosophical dimensions of humanity's deteriorating relationship with nature. Furthermore, it details solution pathways such as nature-based approaches, policy and governance strategies, and education-oriented measures, while offering future scenarios. The chapter concludes with a call for humanity to restore its broken connection with nature.

***Keywords*:** Sand and Dust Storms (SDS), Ecosystem Degradation, Ecological Justice

"A Letter from the Dust"

O human…

I write you this letter from forgotten lakes and vanished valleys.

I am the soil scattered by the wind.

I am the silent scream of a deserted forest, an abandoned field.

Once, I was green.

I embraced the rain upon my body,

I held the wind within my branches.

While I gave you life,

You demanded more and took the life within me.

You plowed, consumed, and forgot.

And now, I am

A cloud of dust rising to the sky—

Blinding your eyes, filling your lungs,

But most of all, shaking your conscience.

You did not only destroy nature and natural resources;

You silenced yourself.

Because I am

The echo of the nature and the resources within you.

I speak to you in the language of dust:

Stop, see, and remember.

The Language of Dust, the Cry of Nature

Sand and dust storms (SDS) are not merely natural phenomena. They are the Earth's silent yet striking cry — a manifestation of the fractures in the balance that humanity has ignored. Each grain of dust carried by these storms bears a message from a disrupted ecosystem, from farmland turned to desert, from a dried-up lake, or from a devastated pasture. Nature speaks, but humankind refuses to listen. Thus, SDS are open letters written by nature as a last resort to those who will not hear.

This chapter approaches SDS not only as physical occurrences but as ecological, philosophical, and existential messages. It examines the relationship between humans and nature, the environmental balances we have disturbed, and the "warning systems" that nature employs to signal distress.

Nature's Reaction: Crossing Ecological Thresholds

For billions of years, the Earth existed as a self-regulating system in balance. Within this system, complex interconnections bound the atmosphere, soil, water, climate, flora (all plant life within a specific region or ecosystem), and fauna (all animal species within a region or ecosystem) together in mutual dependence. Yet today, this integrity has been largely disrupted under the pressure of "human-centered development." Sand and dust storms are among the earliest visible signs of this disturbance.

Deserts constitute the primary sources of dust storms. In the visual below, regions marked in red and yellow — including North Africa, the Middle East, Central Asia, and both South and North America — represent the main dust-producing zones. The scale in the visual indicates that as the value approaches 1, the probability of SDS occurrence increases, while values approaching 0 reflect lower risk levels.

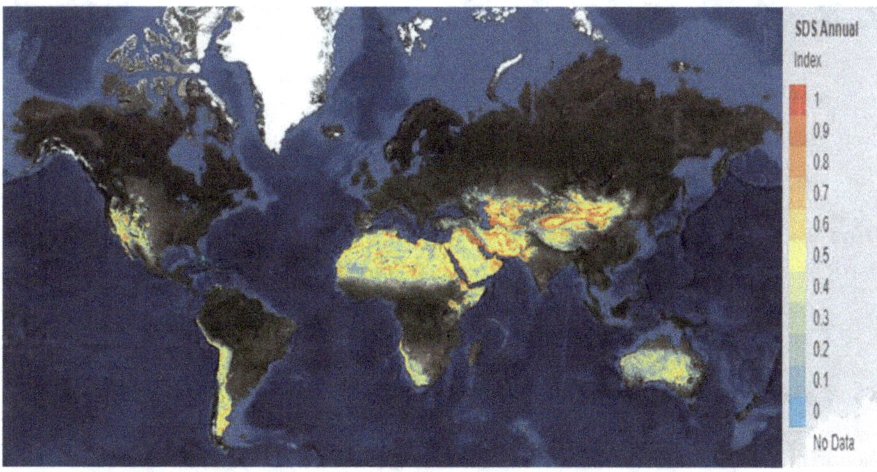

Source: (https://maps.unccd.int/sds/)

Recent studies have demonstrated that the rising frequency and intensity of sand and dust storms are closely associated with anthropogenic drivers such as climate change, prolonged drought, unsustainable land-use practices, and deforestation (Kok et al., 2021; Micklin, 2007). The spatial distribution of these regions, categorized according to predominant land-use types, is illustrated in the figure below. The map was developed collaboratively by the World Meteorological Organization (WMO), the United Nations Environment Programme (UNEP), and the United Nations Convention to Combat Desertification (UNCCD).

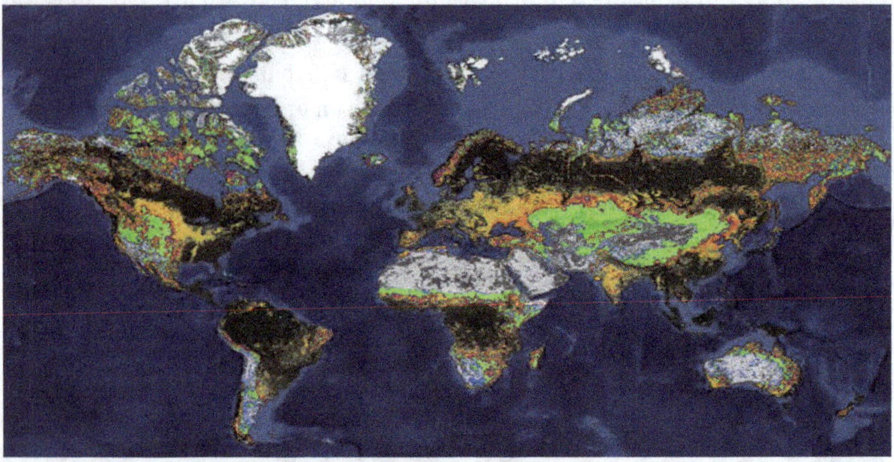

Source: (https://maps.unccd.int/sds/)

Here, nature is not a passive victim but an active responder:

Beyond natural factors, it becomes a producer — a living system that fills the atmosphere with dust as a defensive reaction to human-induced disturbances, particularly the exposure of land surfaces through unsustainable practices.

The Concept of "Threshold" and the Ecological Rupture

According to resilience theory — the capacity of a system to cope with a disturbance or to return to its prior state — every ecosystem can tolerate certain levels of stress. However, once a threshold is crossed, the system undergoes an irreversible transformation (Folke, 2006). Sand and dust storms (SDS) emerge as "feedback mechanisms" triggered by the transgression of these ecological thresholds. The ecosystem, under increasing anthropogenic pressure, responds through a kind of ecological backlash. In this context, SDS are no longer merely physical phenomena; they represent nature's form of resistance — an act of defiance by the Earth and its resources.

The Memory of Dust: Forgotten Balances in Ecosystems

Every grain of sand and dust has been torn from somewhere — from the bed of a dried lake, from an overgrazed pasture, or from the expanding edge of a desert. Each particle bears witness to a place, a time, and a vanishing. The memory of the soil drifts on the wind. Thus, sand and dust storms are not only expressions of nature's violence, but also embodiments of its memory.

For instance, the Aral Sea Basin stands as one of the greatest human-induced environmental catastrophes of the twentieth century. Due to misguided irrigation policies, the lake has lost nearly 90% of its volume, and the exposed areas have become vast sources of dust storms (Glantz, 1999). In this region, nature tells the story of its mistreatment through the clouds of dust it sends across Central Asia each spring. Similarly, the number of dust events originating from China's Gobi Desert has more than doubled — from about 10 occurrences per year in 1990 to over 24 in 2020 (Wang et al., 2021).

Globalized Risks: Nature's Rebellion Becomes an International Language

Sand and dust storms have evolved beyond local phenomena to become global environmental challenges. Dust particles carried by winds traverse continents, altering atmospheric chemistry and influencing plankton production upon reaching the oceans (Yu et al., 2015). While these storms recognize no physical borders, human politics — confined by regional and national boundaries — remains insufficient in addressing their complex transboundary impacts.

Recognizing this escalating threat, the United Nations has declared 2025–2034 as the Decade for Combating Sand and Dust Storms (UNCCD, 2025). Highlighting the severe global implications of SDS across multiple dimensions, the UN General Assembly has also designated July 12 as the International Day of Combating Sand and Dust Storms (A/RES/77/294). Furthermore, during the 16th

Conference of the Parties (COP16) to the United Nations Convention to Combat Desertification (UNCCD), held in Riyadh, Saudi Arabia (2–13 December 2024), Decision 27/COP.16 was adopted, focusing on policy frameworks and thematic follow-up on sand and dust storms.

This declaration represents more than an institutional call — it is a test for humanity: a test of whether we can learn to read and respond to the messages nature continues to send.

Sand and Dust Storms: Causes, Significance, Impacts, Measures, and Future Scenarios

Sand and dust storms constitute a transboundary, multi-factor environmental problem arising from the interaction between land degradation in arid and semi-arid regions, climate change, improper land use, and natural processes. They exert global environmental, social, and economic impacts. Each year, approximately 2 billion tons of dust are lifted into the atmosphere, and these events are becoming increasingly frequent, intense, and widespread (Ginoux et al., 2012; Goudie, 2014; Middleton, 2017).

Addressing the challenge of SDS is beyond the capacity of any single nation; it demands multilateral, coordinated, and effective international action. Globally, SDS originate from about 45 source countries and affect more than 150 countries worldwide (UNCCD, 2022; Aydoğdu & Aydınoğlu, 2025). The quantities of dust generated annually (in tons per year) across continents and regions are illustrated in the figure below.

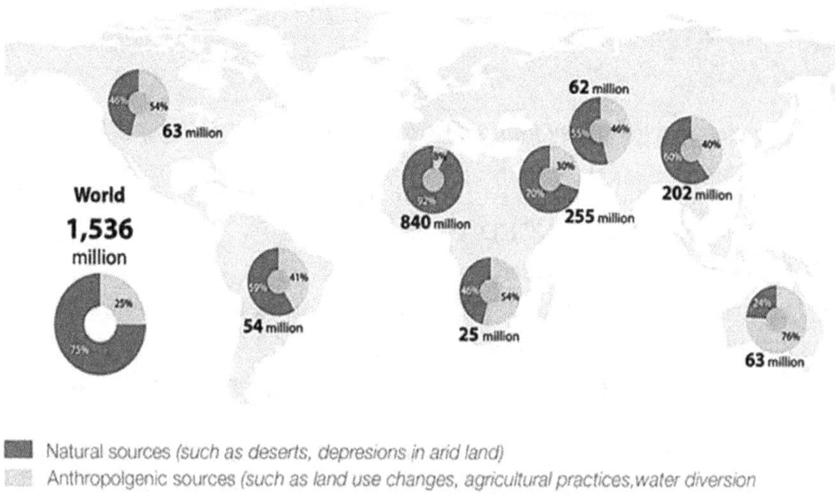

Source: https://maps.unccd.int/sds/

Approximately 330 million people are directly or indirectly affected by sand and dust storms (SDS). Around 2 billion people live in areas with a high risk of exposure to such events. Therefore, addressing the problem requires not only national-level action but also multilateral international cooperation, policy coordination, and the development of early warning systems (WMO, 2020).

Causes of Sand and Dust Storms (SDS)

The factors leading to SDS can be classified into natural and anthropogenic (human-induced) categories.

Natural Factors (accounting for approximately 75–81% of influence)

- Arid and semi-arid climatic conditions combined with low precipitation regimes
- Soil characteristics (e.g., loose texture, fine-grained, and low organic matter content)

- Wind regimes (high wind speeds exceeding the critical threshold of 5–10 m/s) (Goudie, 2014)

Anthropogenic Factors (accounting for approximately 19–25% of influence)

- Land degradation and desertification: billions of tons of fertile soil are lost globally each year (UNCCD, 2022).
- Overgrazing and deforestation
- Unsustainable agricultural practices (such as stubble burning, fallowing, and monoculture cropping)
- Depletion of water resources (e.g., drying of wetlands and excessive irrigation)
- Climate change: rising temperatures and increased drought frequency intensify SDS occurrences (IPCC, 2021).

Significance of Sand and Dust Storms

The importance of SDS can be grouped under three main domains: impacts on ecosystems, global environmental health, and the climate system.

Impacts on Ecosystems:

- SDS damage vegetation cover and reduce soil fertility.
- Long-range dust transport can also contribute positively to certain ecosystems by delivering minerals to oceans and forests (Jickells et al., 2005).

Impacts on Global Environmental Health:

- Particulate matter (PM_{10} and $PM_{2.5}$) can cause severe respiratory diseases.
- In 2019, global dust-related air pollution was reported to cause thousands of premature deaths (WHO, 2021).

Impacts on the Climate System:

- By reflecting solar radiation, dust particles can temporarily slow global warming.
- However, dust deposition on glaciers accelerates melting, contributing to sea-level rise (Painter et al., 2010).

Impacts of SDS

The impacts of SDS can also be analyzed under social and economic dimensions.

Social Impacts:

- *Health:* SDS increase incidences of asthma, COPD, and cardiovascular diseases. In countries such as Iran, Iraq, Saudi Arabia, and Türkiye, hospital emergency admissions rise by 20–40% during dusty days (Bayram et al., 2020).
- *Migration:* Declines in agricultural productivity drive rural migration and social unrest (UNEP, 2016).

Economic Impacts:

- *Agriculture:* China experiences annual agricultural losses of 2–3 billion USD due to SDS (Zhang et al., 2020).
- *Transportation:* In 2022, India recorded 74 train cancellations and 120 flight delays caused by SDS events.
- *Cleaning and maintenance costs:* In major Saudi Arabian cities, annual cleanup expenses reach approximately 300 million USD (FAO, 2021).
- *Energy production:* Solar panel efficiency can decrease by up to 30% due to dust accumulation (Sarver et al., 2013).

Measures for Combating SDS

Preventive and adaptive measures can be categorized as direct actions, policy and governance measures, and education and awareness initiatives.

Direct Measures:

- Establishment of windbreak shelterbelts (afforestation)
- Soil stabilization (through organic matter enhancement, mulching, or application of binding agents on desert surfaces)
- Protection and restoration of wetlands

Policy and Governance Measures:

- International cooperation (e.g., UNCCD and WMO SDS-WAS programs)
- Early warning systems, particularly those developed by the World Meteorological Organization (WMO)
- Participatory land management: involvement of local communities significantly enhances the effectiveness of mitigation efforts.

Education and Awareness:

- Raising public awareness at the local level
- Encouraging farmers to adopt sustainable agricultural practices

Future Scenarios

Linkages with Global Climate Change:

- According to the IPCC RCP 8.5 scenario, dust-related events are projected to increase by 25–35% by the year 2100 (IPCC, 2021).

Most Vulnerable Regions:

- Middle East and North Africa (MENA): regions facing severe water stress and extensive land degradation.
- Central Asia: the Aral Sea Basin and China's Gobi Desert are among the most dramatic examples.
- Türkiye: Southeastern Anatolia, the Konya Basin, and Iğdır are identified as the most vulnerable areas.

Situation in Türkiye

Among the primary factors shaping the future trajectory of sand and dust storms (SDS) in Türkiye are global climate change, land degradation, water stress, and the condition of vegetation cover. Countries situated within semi-arid climate zones, such as Türkiye, are disproportionately affected by these dynamics.

According to data from the Turkish State Meteorological Service (MGM), 106 dust transport events were recorded across Türkiye in 2022—a 70% increase compared to the 1990s (MGM, 2023).

The IPCC Sixth Assessment Report (AR6) projects that under the high-emission scenario (RCP 8.5), temperature increases across much of Türkiye may reach 4–6°C by 2100. This rise will lead to a 20–30% decline in soil moisture, an increase in evaporation, and consequently, an expansion of dust-prone surfaces (IPCC, 2021).

Particularly in regions such as Southeastern Anatolia, the Konya Closed Basin, the depression zones of Eastern Anatolia, and the Iğdır Plain, both the weakening of soil cover and reductions in agricultural productivity are expected to result in a 30–45% increase in SDS events (Demirtaş et al., 2022). According to MGM's (2023) projections, after 2050, the number of dusty days in Türkiye's central and southeastern regions could exceed 120 days per year.

Furthermore, regional climate modeling studies using RegCM4 and PRECIS indicate that during 2071–2100, Türkiye will experience increased wind speeds during spring and summer, raising dust transport potential by 20–35% (Öztürk et al., 2018).

These scenarios depict not only a climatic phenomenon but also a complex risk landscape encompassing public health, food security, energy efficiency, and migration. Consequently, Türkiye must urgently restructure both its climate adaptation policies and land management frameworks. In the future, nature-based solutions, land rehabilitation, and regional early warning systems will play a critical role in mitigating SDS risks.

Reestablishing Dialogue with Nature: Harmonious Lifestyles and Restorative Ecology

Modern civilization refers to nature merely as a resource; yet nature is not just a resource — it is a being, a subject with agency. The most effective strategies to prevent sand and dust storms are not those developed against nature, but those developed with nature and in harmony with it. Sand and dust storms are not merely environmental phenomena — they are multidimensional crises with economic, social, political, health, and security dimensions. Climate change has further intensified these dynamics. As global temperature rise surpasses 1.1°C, the world's most vulnerable regions — such as the Sahel, the Aral Basin, and the Middle East — are already experiencing warming beyond 2°C (IPCC, 2023). In these areas, vegetation loss, declining soil moisture, and increased evaporation directly contribute to the rising frequency of SDS events (Shao et al., 2011).

Since the early 2000s, the frequency of SDS has increased by over 30% in North Africa and Central Asia (Goudie, 2014; Middleton, 2017). Long-term monitoring in China's Loess Plateau reveals that SDS-related agricultural yield losses can reach up to 15% annually for crops such as wheat (Journal of Arid Environments, 2020). In the Middle East, countries such as Iran and Iraq experience air quality index (AQI) levels exceeding WHO thresholds for more than 100 days each year, leading to 20–35% increases in respiratory diseases (WHO, 2018). In the coming decades, escalating climate change and land degradation are expected to further intensify these events. Yet all these data point to a deeper truth: the root of the problem lies not solely in nature itself, but in the way humanity relates to nature. As long as we regard soil merely as a production area, forests as timber sources, and the atmosphere as a waste repository, nature will continue to respond in its own language — with sand, dust, storms, and drought.

Hence, multi-stakeholder, multi-scale, participatory, and nature-based approaches are more necessary than ever. Nature-based solutions — such as afforestation with native species, windbreak establishment, soil conservation practices, and the integration of traditional knowledge with modern science — exemplify collaboration with nature (FAO, 2022).

Dust in the Shadow of Ecological Justice

The impacts of SDS are not distributed equally across geographies. The poor, rural, and marginalized communities face the harshest consequences. From the perspective of ecological justice, it is deeply unjust that those least responsible for environmental degradation are the most affected by its consequences (Middleton, 2017).

According to World Bank (2021) data, in regions exposed to SDS, health expenditures of low-income populations increase by 20–40%, while household incomes decline by up to 20% due to labor losses. These inequalities are not merely ecological — they are also moral. Those who harm nature the least are often the ones most burdened by its retaliation.

Chapter Conclusion

The Message beneath the Dust: Sand and dust storms are no longer just natural events; they are the visible and transportable expressions of ecological collapse. This dust does not only obscure our sight — it clouds our ability to perceive reality. SDS are part not only of the physical but also the ethical dimension of nature. When humanity neglects the environment, natural resources, and ecosystems, nature responds — in its own language: sometimes as wind, sometimes as floods, sometimes as drought, and sometimes as vast layers of dust covering the sky. We must now reconsider not only the storms — but our own path.

This dust is not merely the residue of the earth — it is the reflection of humanity's loss of direction. Nature and its resources still offer us a chance — but this chance is not infinite.

- If sustainable land use does not become widespread, the anger of the soil will turn into storms.
- If all actors fail to take responsibility in climate policy, the winds will carry more dust each year.
- If nature-based solutions are not implemented, it is we who will continue to disintegrate.

These storms are signals of ecological degradation — reminders that humanity has severed its connection with nature. It is time not for conflict with nature, but for a renewed dialogue in harmony with it.

"Nature does not take revenge. It reminds."— An anonymous grain of dust

It is time to warm not only our climate, but also our conscience. We failed to understand that the wind is a warning, the soil a prayer, and the tree an embodiment of patience. Yet it is not too late to act. We must no longer seek to dominate nature, the environment, and natural resources — but to walk alongside them, in silence, listening.

"The tears of the earth are carried by the wind; because humanity refuses to hear its silence." — A shadow of the desert

In Short: We Either Return to Nature — or to Dust

We must now reconsider not only the storms but also our path forward. What has been expressed in this chapter is not merely an analysis, but also a call. Nature still wishes to see humanity as a friend — but to earn that, humanity must once again become natural.

Chapter References

Atabey, E. (2024). Çöl Tozları / Toz Fırtınaları ve Sağlığa Etkileri *(Desert Dusts / Dust Storms and Their Effects on Human Health).* Retrieved from https://www.researchgate.net/publication/377738747_COL_TOZLARI_TOZ_FIRTINALARI_VE_SAGLIGA_ETKILERI

Aydoğdu, M. H., & Aydınoğlu, E. (2025). Kum ve Toz Fırtınalarının Sosyal, Siyasal ve Ekonomik Boyutlarına Genel Bir Bakış ve Önlem Stratejileri Önerileri *(A General Overview of the Social, Political, and Economic Dimensions of Sand and Dust Storms and Proposed Preventive Strategy Suggestions). Latin America 10th International Conference on Scientific Researches*, May 1–4, Havana/Cuba, s. 458–464. ISBN: 978-625-5962-70-6

Azorin-Molina, C., Asin, J., McVicar, T. R., Minola, L., Lopez-Moreno, J. I., Vicente-Serrano, S. M., & Chen, D. (2018). Evaluating anemometer drift: A statistical approach to correct biases in wind speed measurement. *Atmospheric Research*, 203, 175–188.

FAO. (2021). *Sand and Dust Storms in the Middle East and North Africa (MENA): A Regional Overview.* Food and Agriculture Organization of the United Nations. Retrieved from https://documents1.worldbank.org/curated/en/483941576489819272/pdf/SAND-AND-DUST-STORMS-IN-THE-MIDDLE-EAST-AND-NORTH-AFRICA-MENA-REGION-SOURCES-COSTS-AND-SOLUTIONS.pdf

FAO. (2022). *The State of the World's Land and Water Resources for Food and Agriculture (SOLAW).* Rome: FAO. https://doi.org/10.4060/cb9910en

Folke, C. (2006). Resilience: The emergence of a perspective for social–ecological systems analyses. *Global Environmental Change*, 16(3), 253–267.

Ginoux, P., Prospero, J. M., Gill, T. E., Hsu, N. C., & Zhao, M. (2012). Global-scale attribution of anthropogenic and natural dust sources and their emission rates. *Atmospheric Chemistry and Physics*, 12(7), 3037–3056.

Glantz, M. H. (1999). *Creeping Environmental Problems and Sustainable Development in the Aral Sea Basin.* Cambridge University Press, UK.

Goudie, A. S. (2014). Desert dust and human health disorders. *Environment International*, 63, 101–113.

IPCC. (2021). *Climate Change 2021: The Physical Science Basis.* Contribution of Working Group I to the Sixth Assessment Report of the Intergovernmental Panel on Climate Change. Cambridge University Press. Retrieved from https://www.ipcc.ch/report/ar6/wg1

IPCC. (2023). *Climate Change 2023: Synthesis Report.* Intergovernmental Panel on Climate Change. Retrieved from https://www.ipcc.ch/report/ar6/syr/

Jickells, T. D., et al. (2005). Global iron connections between desert dust, ocean biogeochemistry, and climate. *Science*, 308(5718), 67–71.

Journal of Arid Environments. (2020). Impacts of dust storms on agricultural productivity in the Loess Plateau, China. *Journal of Arid Environments*, 178, 104154. https://doi.org/10.1016/j.jaridenv.2020.104154

Kok, J. F., et al. (2021). Toward a greater understanding of the atmospheric dust cycle. *Science*, 367(6480), eaaz9507. https://doi.org/10.1126/science.aaz9507

MGM. (2023). *Türkiye'de Toz Taşınımı Projeksiyon Raporu* (*Dust Transport Projection Report in Turkey*). Ankara: Meteoroloji Genel Müdürlüğü Yayınları. Retrieved from https://www.mgm.gov.tr/

Micklin, P. (2007). The Aral Sea disaster. *Annual Review of Earth and Planetary Sciences, 35*(1), 47–72.

Middleton, N. J. (2017). Desert dust hazards: A global review. *Aeolian Research, 24*, 53–63. https://doi.org/10.1016/j.aeolia.2016.12.001

Oğuz, K., Omak, İ., Dündar, C., & Coşkun, M. (2017). *Hava, Toz ve İklim Merkezi (Weather, Dust and Climate Center - WDCC)* (*Air, Dust and Climate Center*). ResearchGate. Retrieved from https://www.researchgate.net/publication/318340933

Oktay, M. M., Al, B., Boğan, M., Kul, S., Gümüşboğa, H., & Bayram, H. (2023). Impact of desert dust storms, PM10 levels and daily temperature on mortality and emergency department visits due to stroke. *Frontiers in Public Health, 14*(11), 1218942. https://doi.org/10.3389/fpubh.2023.1218942

Özdemir, E. T., Türker, Y., Kolay, O., & Tekin, F. B. (2021). *Habub Nedir? Örnek Olay İncelemesi: 12 Eylül 2020, Polatlı, Ankara* (*What is a Haboob? A Case Study: 12 September 2020, Polatlı, Ankara*). Retrieved from ResearchGate. https://www.researchgate.net/publication/352550569

Painter, T. H., et al. (2010). Impact of disturbed desert soils on duration of mountain snow cover. *Geophysical Research Letters, 37*(12). https://doi.org/10.1029/2009GL041585

Rezaei, R., Güllü, G., & Ünal, A. (2025). Assessing the impact of climate change on summertime tropospheric ozone in the Eastern Mediterranean: Insights from meteorological and air quality modeling. *Atmospheric Environment, 344*(18), 121036. https://doi.org/10.1016/j.atmosenv.2025.121036

Rockström, J., et al. (2009). A safe operating space for humanity. *Nature, 461*(7263), 472–475. https://doi.org/10.1038/461472a

Sarver, T., Al-Qaraghuli, A., & Kazmerski, L. L. (2013). A comprehensive review of the impact of dust on the use of solar energy: History, investigations, results, literature, and mitigation approaches. *Renewable and Sustainable Energy Reviews, 22*, 698–733.

Shao, Y., et al. (2011). Dust cycle: An emerging core theme in Earth system science. *Aeolian Research, 2*(4), 181–204.

Shiva, V. (2005). *Earth Democracy: Justice, Sustainability, and Peace.* South End Press. Retrieved from https://www.amazon.com/Earth-Democracy-Justice-Sustainability-Peace/dp/089608745X

Steffen, W., Crutzen, P. J., & McNeill, J. R. (2007). The Anthropocene: Are humans now overwhelming the great forces of nature? *AMBIO: A Journal of the Human Environment, 36*(8), 614–621.

UNCCD. (2022). *Global Land Outlook (2nd ed.).* Bonn: UNCCD Secretariat. Retrieved from https://www.unccd.int/resources/global-land-outlook

UNCCD. (2025). *Sand and Dust Storms Source Base-map.* Retrieved from Retrieved from https://maps.unccd.int/sds/

UNEP. (2016). *Sand and Dust Storms: A Global Environmental Issue.* United Nations Environment Programme. Retrieved from https://wesr.unep.org/media/docs/assessments/global_assessment_of_sand_and_dust_stormsx.pdf

Wang, X., et al. (2021). Recent intensification of dust storm activity over East Asia. *Atmospheric Chemistry and Physics, 21*(7), 5615–5633.

WHO. (2018). *Ambient Air Pollution: A Global Assessment of Exposure and Burden of Disease.* Geneva: World Health Organization. Retrieved from https://iris.who.int/bitstream/handle/10665/250141/9789241511353-eng.pdf

WHO. (2021). *Air pollution and health.* World Health Organization. Retrieved from https://www.who.int/news-room/fact-sheets/detail/ambient-(outdoor)-air-quality-and-health

WMO. (2020). *Sand and Dust Storm Warning Advisory and Assessment System (SDS-WAS): Annual Report.* World Meteorological Organization. Retrieved from https://community.wmo.int/en/activity-areas/gaw/science-for-services/sds-was

World Bank. (2021). *The Cost of Air Pollution in the Middle East and North Africa.* Washington, D.C.: World Bank. Retrieved from https://openknowledge.worldbank.org/handle/10986/35035

Yu, H., et al. (2015). Quantification of trans-Atlantic dust transport from satellite observations. *Atmospheric Chemistry and Physics, 15*(5), 2787–2807.

Zhang, X. Y., et al. (2020). Dust storm damages and economic losses in China: Status and challenges. *Natural Hazards, 103*, 1183–1197.

Back Cover Text

Nature, Humanity, and the Fragile Balance of Existence (*A Philosophical Inquiry into the Ethical, Ecological, and Existential Dimensions of Nature*) invites readers on a profound journey through the intertwined destinies of the Earth and humankind.

From the sacred origins of nature in faith and philosophy to the scientific realities of climate change, water scarcity, soil erosion, and desertification, this book explores how our moral choices shape the planet's fate. It is a multidisciplinary reflection on ethics, ecology, and existence — where theology meets data, and where philosophy meets policy.

Through a holistic lens that bridges global and local perspectives, it questions humanity's role in the age of crisis: *Are we the guardians or the violators of the Earth's harmony?*

With insights that blend scientific precision and moral depth, **Mustafa H. Aydoğdu** calls for a new ecological conscience — one rooted in justice, moderation, and reverence for life itself.

To restore balance, we must first remember what it means to belong to nature.

www.ingramcontent.com/pod-product-compliance
Lightning Source LLC
LaVergne TN
LVHW052307200926
848481LV00002B/494
9798899663482